FUNDAMENTALS OF SOLAR HEATING

Prepared by
Sheet Metal and Air Conditioning
Contractors National Association

For The
U.S. DEPARTMENT OF ENERGY
Assistant Secretary for Conservation
and Solar Applications

January 1978

University Press of the Pacific
Honolulu, Hawaii

Fundamentals of Solar Heating

Prepared by

Sheet Metal and Air Conditioning Contractors National Association

For The

U.S. Department of Energy

Assistant Secretary for Conservation and Solar Applications

ISBN 0-89875-089-X

University Press of the Pacific

Honolulu, Hawaii

http://www.UniversityPressofthePacific.com

ACKNOWLEDGEMENTS

This training course in *Fundamentals of Solar Heating* was made possible by a grant from U.S. Department of Energy (DOE) Solar Technology Transfer Program.

The essential purpose of the training course is to provide the on-the-job, air-conditioning industry employee with the opportunity to learn about solar technology through supervised correspondence instruction.

The program is a joint effort of the Sheet Metal and Air Conditioning Contractors' National Association (SMACNA) and the Northamerican Heating and Air-Conditioning Wholesalers Association (NHAW).

Technical data and illustrations for this course were secured from a number of sources. Nearly 100 manufacturers contributed diagrams, photographs, and proprietary instructional materials. The Solar Energy Applications Laboratory, Colorado State University, provided materials used in their one week training program held regularly at CSU. Information from DOE and HUD publications were also used extensively as well as materials from courses offered by NHAW's accredited Home Study Institute.

All of this material was structured into a workable home study format by a team of educators, engineers, and comfort-heating personnel directed by Dr. James J. Buffer, Jr., Professor in the Academic Faculty of Industrial Technology, The Ohio State University (Columbus, Ohio). Other major participants in this project included James Healy (NHAW); and Dr. William D. Umstattd, Dr. E. Keith Blankenbaker, and James A. O'Brien of The Ohio State University.

Solar Heating is in a stage of dynamic growth which is fostering many innovative solar heating applications. No single treatment of this expanding technology could include all noteworthy developments. However, this training course does provide the essential information that a practicing heating and air-conditioning wholesaler, contractor, and technician needs to advance in *sizing, installing,* and *servicing* practices as the market for solar heating progresses.

TABLE OF CONTENTS

Table of Contents
(continued)

Table of Contents
(continued)

INTRODUCTION

HEAT FROM THE SKY

One estimate made by engineers and scientists is that one-fourth of all the energy used in the United States is consumed in heating buildings. Billions of Btu's from coal, oil, gas, electricity are consumed every heating season. We are depleting our supply of readily useable fuels at an ever-increasing rate. Fortunately, future generations will have an energy source which, every three days, equals the combined energy in all the available coal, oil, and gas in the world. That energy source is the *sun*.

This amount of potentially useful energy is staggering, but logical. Since all forms of energy originate from the sun, the basic source is, by necessity, a reservoir of almost unimaginable size.

A large quantity of energy is needed for heating and a large amount of energy is produced by the sun. These two facts, coupled with: (1) the large quantity of energy needed for heating and (2) the large amount of energy emitted by the sun—when coupled with a growing need to conserve fossil fuels—make solar house heating a desirable and promising area for research toward easing the demands on the world's resources.

Until a few years ago, there was little, if any, general interest in solar applications to heating and cooling. Today, however, there are many solar energy research and development activities in the heating and air-conditioning industry.

Solar assisted heating is not a new idea. Solar heaters were used in the Army camps of Southern California during World War I, and solar water heaters received at least some use in some southern states during the 1930's.

But, prior to 1972, solar energy research applied to this industry was essentially privately funded. Now, federal funds are being allocated for solar programs and have had an impact on research and development activities.

A joint National Science Foundation—National Aeronautical and Space Administration report published in 1972 is generally credited with having had the most public and political impact on bringing renewed attention to the potentials of solar energy. The report concluded:

1. Solar energy is received in sufficient quantity to make a major contribution to future U.S. heat and power needs.
2. There are no technical barriers to wide application of solar energy to meet U.S. needs.

With this endorsement and others, Congress passed the ERDA Bill authorizing the establishment of the Energy Research and Development Administration. Among other things, ERDA replaced the old Atomic Energy Commission. On October 1, 1977, the Department of Energy (DOE) assumed the overall responsibility regarding energy.

Compared to the meager funding that was obtained by a few solar energy pioneers, there is now relatively substantial federal money earmarked for solar energy research, development, and demonstration projects.

FEDERAL GOALS

The present overall goal of the federal programs for heating and cooling is to stimulate industrial and commercial capability (including that of small business) to produce and distribute solar heating and cooling systems and, through widespread applications, reduce the demand on present fuel supplies.

If they are successful, government efforts will stimulate the private sector to support the incorporation and installation of combined solar heating and cooling systems in existing buildings in the United States. For the period beyond 1970, the following are ideal goals for new construction and retrofitting in existing building:

By 1980: Incorporate solar energy systems in at least 10% of the annual residential and commercial building starts and the installation of retrofit systems annually on 2,500 residential and 200 commercial buildings.

By 1985: Incorporate solar energy systems in at least 10% of annual residential and commercial building starts and the installation of retrofit systems annually on 25,000 residential and 1,000 commercial buildings.

Perhaps some useful perspective can be gained by noting a similarity between the status of solar assisted heating *today* and electric heating in the *late 1950's*. Then, electric interests (generally people outside the traditional heating-cooling industry) began to promote and market electric heating systems. Today, there are solar interests who want to market solar systems. Electric heat in the 50's was unbelievably expensive compared to gas and oil-fired central systems.

Like today's solar assisted systems, the market for electric heat in 1959 appeared (to "traditional" heating men) to have very little potential. Only certain types of buildings—those with many lights and heavy insulation—were considered "suitable" for electric heat. Today, solar systems are also said to be suitable usually only in special situations.

It is currently agreed that electric heat applications have grown beyond the expectations of fifteen years ago when few persons foresaw today's fuel costs or shortages.

In the 1950's, an important "marriage" did begin to take place. Names familiar to the heating technicians began manufacturing electric heating products—and the electric heat interests became part of the heating-cooling industry. Electric heat began to be marketed through the people who *know* heating and air-conditioning. The over-zealous enthusiasts who saw electric heat as something magical, that did *not* have to meet traditional comfort design criteria, were squeezed out. Also, there were numerous front page newspaper stories about electric heating installations misapplied and "ripped-out" by irate homeowners.

AN IMPORTANT LESSON?

The evolution of electric heating, therefore, may serve as an important history lesson for both solar enthusiasts and traditional heating-cooling practitioners.

Solar interests should not try to do it alone.

Heating technicians *know* the heating business. Nor should the solar interests believe, as the electric interests once did, that there is something magical about the "fuel source" that puts solar assisted heating above the technology that has evolved since the turn of the century.

Also, solar interests should avoid customer oversell. In an age of consumerism, the heating industry does not need the stigma of "scare" headlines regarding energy shortages or poorly designed solar systems. Either may cause the homeowners to "tear-out" thousands of dollars worth of solar equipment because of extreme product dissatisfaction.

And for those of us who are a part of the heating industry, just as electric heat was not for everybody in '59—and it may *still* not be part of everyone's business today—solar assisted heating and cooling may *never* be apart of *everyone's* business tomorrow.

But, heating specialists should not resist solar assisted heating and cooling because of lack of technical information. Let's make an honest effort to learn what the solar specialists has to offer, and then teach them about comfort conditioning practices and how to provide good consumer products.

James Healy
Director of Education
Northamerican Heating and Air-
conditioning Wholesalers Assn.

SOLAR HEATING AND COOLING

Solar energy is rapidly becoming a logical alternative source of heat as the cost and unavailability of conventional fuels becomes a major problem in industrial countries. Getting heat from the sun is not a new idea—most people have all experienced getting sunburned on a cloudy day, much to their surprise, so the energy is there. And now, technology has brought the cost of harnessing the sun closer to being economically competitive. Add the fact that solar heating and solar cooling are very attractive environmentally, and the reasons for being involved in this course are almost complete. This course will provide fundamental technical knowledge on solar assisted heating. There will be a demand for technicians with an understanding of the design, installation, and servicing responsibilities as the market for solar assisted heating develops. The question "Why am I here?" is now answered. The next question is "What is involved?"

PASSIVE SOLAR HEATING SYSTEM

In its simplest terms, a solar heating/solar cooling system is any system which reduces consumption of conventional fuels by utilizing the sun's energy as a method of heating or cooling. The system can be either *passive* or *active.* People utilize passive solar energy when opening the curtains on the sunny side of the house during the cold months to let the sun add its warmth. There is a minor amount of control over this system during the summer when the curtains are closed to block out the sun's rays and the unwanted heat radiated within the house. Thus, it is fairly safe to state that a passive solar energy system has no moving parts, except of course for the opening or closing of a window shade or curtain, and is mainly the concern of the building designer and/or the

landscape architect. It is also safe to conclude that this type of solar energy is only incidental to this course and will be discussed only when it benefits the primary point of this course, the *controlled collection* and *distribution* of solar heat. Some examples of passive systems are depicted below:

An early discussion of designing a passive solar system which includes taking advantage of the sun by such means as window orientation and roof overhang to conserve energy can be found in the University of Illinois Small Homes Council circular which was published around 1945.

Active Solar Heating System

The alternative to a passive system is an active system. To understand and visualize what goes into an active solar heating/solar cooling system, the following schematic illustration is presented in Figure 1-4.

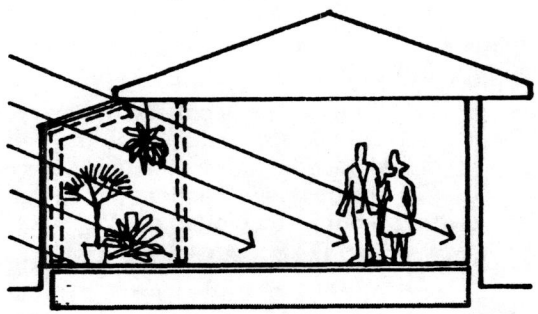

Fig. 1-2. Greenhouse as a Solar Collector. Broken lines note movable insulation.

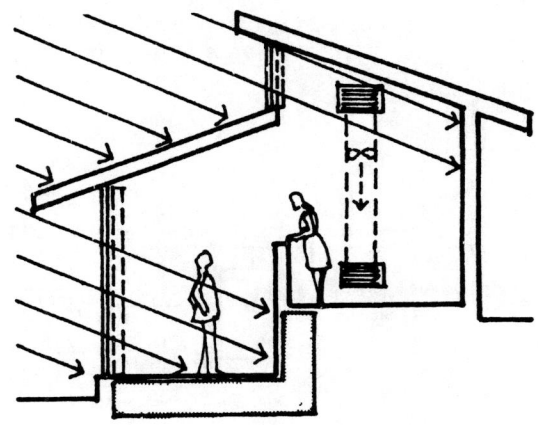

Fig. 1-3. Roof Monitor as a Solar Collector. Duct and fan circulates trapped hot air back to floor level.

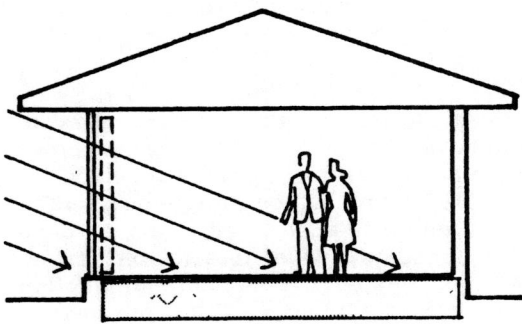

Fig. 1-1. South facing window.

While this illustration shows ten parts to a total system, only those items in the illustration which are represented by rectangular blocks will be discussed in this lesson. The other four items—pumps, blowers, automatic valves, and dampers—are important to the system and will be dealt with later in this course.

Active systems currently consist of the following six essential units to collect, control, and distribute solar heat.

Unit	Function
1. Solar Collector (Fig-1)	Intercepts solar radiation and converts it to heat for transfer to a thermal storage unit or to the heating load.

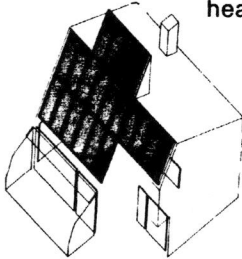

2. Thermal Storage Unit	Can be either an air or liquid unit.
(rock storage shown)	If more heat is collected than required for space heating or domestic water heating, then the heat is stored for later use at night or on cloudy days.

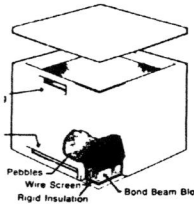

Can be either a liquid, rock, or phase change unit. Rock storage for an air unit is shown in Figure 2. Storage units generally operate independent of any cooling or heating requirement since they can be collecting and storing solar energy whenever

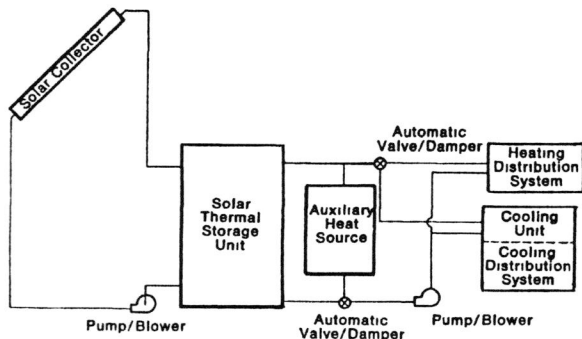

Fig. 1-4. Schematic Representation of a Solar Heating and Cooling System.

there is sufficient solar radiation.

3. Auxiliary Heat Source	Used to back up heat demands when the solar/collector thermal storage system is unable to meet the demands; such as periods of high heating/cooling load demands and low solar availability.

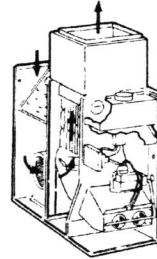

While it is possible to design and construct a solar system that could meet total needs, it is generally not as economical (cost vs. savings) as an auxiliary assisted system.

4. Heat Distribution System and **5. Cooling Distribution System**	Depending on the systems selected for solar heating/solar cooling, these could be the same system or different systems.

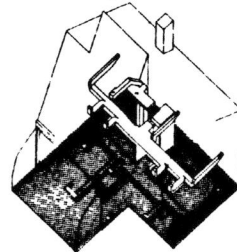

Usually a blower and duct distribution capable of utilizing heat directly from either the solar collector or the thermal storage unit. In some other cases, it can be a liquid system where the liquid is piped directly to the space to be heated/cooled and where fan coil units are used to heat/cool room air.

6. Cooling Unit (absorption unit shown)	The methods used for cooling include absorption cooling units (lithium-bromide or ammonia-water with the former the only commercially available system and even then it has been only used in experimental installations), Rankine cycle vapor-compression, and others. Alternatively, a heat pump might be utilized as a conventional cooling unit (powered by electricity) and used as the auxiliary for the solar space heating system.

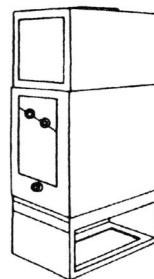

A desirable feature not included above as part of the basic heating/cooling unit is the provision to use solar energy to also heat domestic hot water (DHW). This preheated water requires little added conventional energy (e.g., gas or electricity) to boost its temperature to the desired range of approximately 140°F. (More information on domestic water heating will be provided in subsequent lessons.)

ACTIVE SYSTEM DETAILS

This section of the lesson is designed to provide greater detail about the six major units of a solar space heating/solar cooling system. It describes alternatives to the most common component choices within each part of the system.

Solar Collector Configurations

Solar collectors can be of the following configurations:

Flat Plate Collectors

The construction of the present commercially available flat plate collector consists of an absorber plate (usually a blackened metal surface) which absorbs direct and diffuse solar radiation and converts the energy to heat. The heat in the absorber plate is transferred to an appropriate liquid or air stream which is in contact with the absorber plate and delivers the heat to the next part of the system—the Thermal Storage Unit. Since the absorber plate tends to lose heat (upward through the cover plate, side, and back) the other components of the collector are designed to reduce the loss as much as possible as is seen in the next two illustrations.

The insulation beneath the absorber and the transparent covers reduces heat loss through radiation, conduction, and/or convection. Glass covers do not obstruct the solar radiation passing through, but they do reduce convection loss since any air movement (including wind) across the absorber is absent. Also, the air space between the glass and the absorber acts to reduce conduction loss between these two components.

Concentrating Collectors

Concentrating collectors gather direct solar radiation over a large area and then focus (concentrate) the radiation onto a smaller absorber area. The effect of this concentrating is to increase the fluid temperature delivered from the collector. An interesting point is that the quantity of heat gained is nearly the same as for a flat plate collector with the same aperture area.

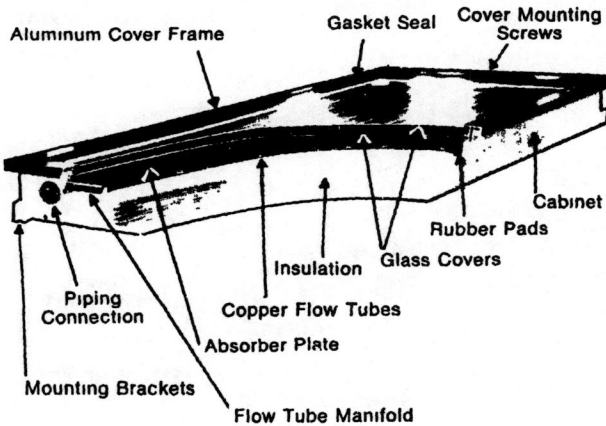

Fig. 1-5. Liquid Collector.

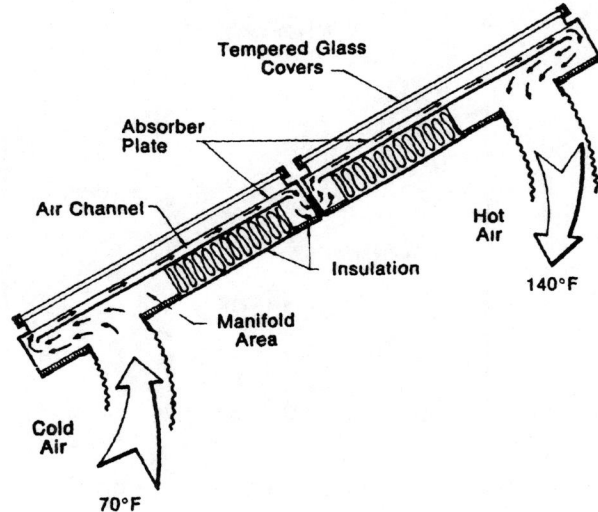

Fig. 1-6. Air Collector.

Solar Collector Configurations		
Collector Type	Absorbs	Heats
Flat Plate	Direct and Diffuse Solar Radiation*	Liquid
Flat Plate	Direct and Diffuse Solar Radiation*	Air
Concentrating	Direct Solar Radiation	Liquid

*See page 1-9 for the difference between diffuse and direct solar radiation.

The major *disadvantages* to the use of concentrating collectors are:

1. Their inability to function on cloudy or overcast days;
2. They must track the sun's travel; and
3. They are expensive to construct, operate, and maintain.

Cloudy days provide only scattered (diffuse) solar radiation. The illustration below shows that concentrating collectors will not work on cloudy days and, thus, they can only operate on clear sunny days simply because diffuse radiation cannot be focused on the absorber by the reflector. However, flat plate collectors can catch and absorb the heat of both *direct* and *diffuse* solar radiation.

Thermal Storage Choices

Collected heat not used immediately is stored by heating a particular substance that can "hold" the heat for later use. While a variety of substances can be used, water and rocks are most popular. Water storage is used with liquid solar collectors; rocks are used with air collectors. Water can be contained in some type of leak-proof "tank" and rocks are enclosed inside some type of storage "bin."

All storage units should be insulated, but water tanks should be particularly well insulated. Insulation prevents loss of stored heat and minimizes the amount of uncontrolled heat escaping from storage that might tend to overheat nearby rooms (similar to a hot furnace room "overheating" adjacent spaces).

Water Storage. Heated water from a solar collector is discharged into the top of a water storage tank. Cooler water at the bottom is then pumped back through the collectors to be reheated.

Water storage units can be placed in a basement, crawl space, garage, or buried inside or outside the structure. Buried units are seldom a good idea because of service problems, damage to insulation, and corrosion. Tanks can be made of watertight concrete, steel or (sometimes) fiberglass. Because water is relatively heavy (8 and 1/3 pounds per gallon), large tanks must be well supported and bottom insulation must not be crushed (thereby reduce insulating effectiveness). Figure 1-10 shows one method to insulate the bottom of a tank.

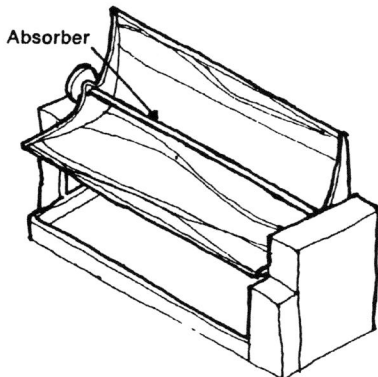

Fig. 1-7. Linear Concentrating Collector.

Fig. 1-8. Circular Concentrating Collector.

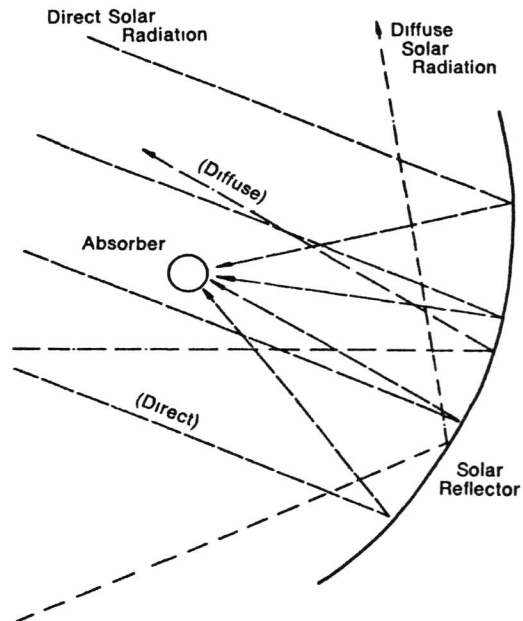

Fig. 1-9. Direct and Diffuse Radiation on a Solar Concentrator.

Air Storage. In air storage units, heated air from the collector is circulated through the rocks to transfer heat.

Rock storage can be provided by assembling wood from boxes, by using concrete blocks, or in cylindrical steel containers. Again, containers must be well constructed to prevent bulging and eventual collapse of the storage container. Although the container may take almost any shape, a basic cube configuration (Figure 1-11) is preferred to configurations that would cause long vertical or extended horizontal air flow paths through the pebble-bed.

However, if the height of the storage unit is limited to less than five feet, a horizontal design can be used. The horizontal design does present a few drawbacks: because of its greater length rather than height as in a cube unit, it imposes a greater pressure drop to air flow. Further, since warm air tends to rise, great care is needed to insure that the air flow goes through the entire length and depth of the horizontal unit. A tight fitting lid with baffle plates which force the air flow down and through the pebble bed helps overcome these problems.

The pebbles in these units should be of uniform size and not fractured to insure adequate air flow. The pebbles should be washed to remove dust before being placed into the unit.

Provision should be made for access around these units for maintenance. The pebble units could suffer structural failure. The water units are subject to leakage, especially at the plumbing connections. Steel tanks are subject to corrosion so space should be provided for access if replacement is necessary.

Auxiliary Heat Sources

With current energy shortages greatly discouraging the use of petroleum products or by-products (oil, LPG, natural gas), the trend is toward electrical space heating systems as back-ups for solar heating systems. The two types most often used are *electric resistance furnace* heating and *heat pumps,* with additional electric resistance heating usually serving as the back-up to the heat pump.

Since all the data up to this point have dealt with a solar heating system, the following two illustrations show a solar air and a solar water system. When comparing the two systems, take the following items into consideration.

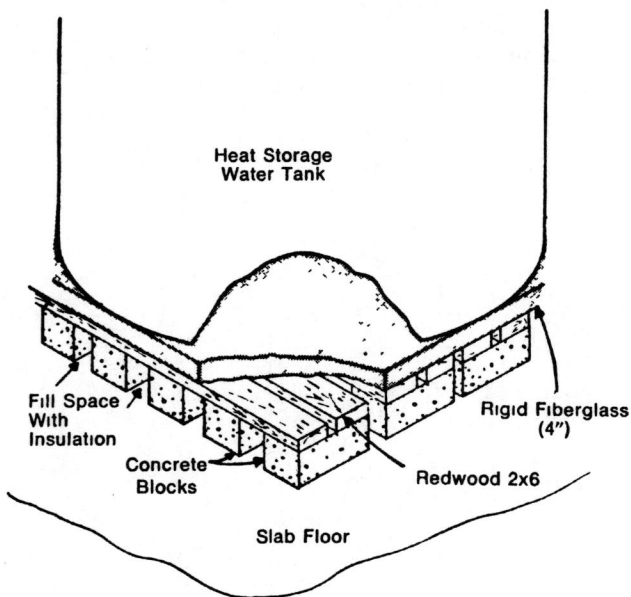

Fig. 1-10. Bottom Insulation and Support for Water Storage Tanks.

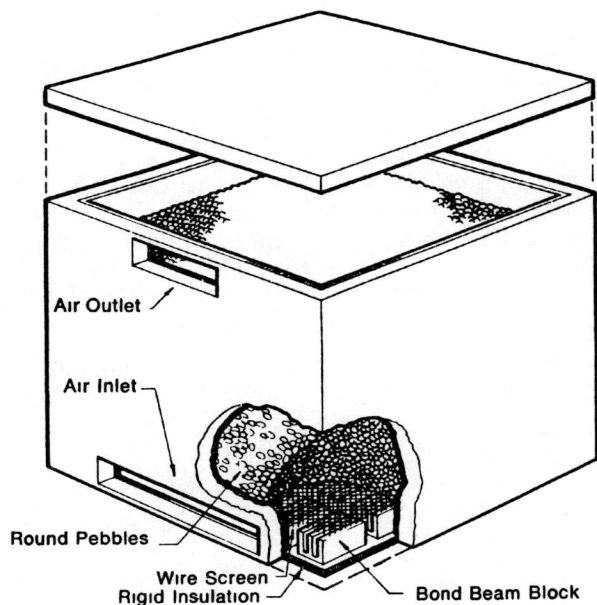

Fig. 1-11. Pebble-Bed Heat Storage Unit.

Warm Air System Fig. 1-12(A)

Plus	Minus
Lower capital costs.	Duct work risers occupy usable floor space.
No corrosion, clogging, or anti-freeze solutions.	More energy required to move solar energy from collector to storage to use.
Leaks have less severe consequences.	Installation must be leak tight which is less obvious than with water systems.
Domestic hot water system is not subject to contamination by leaks from the heat storage unit.	

Warm Air System Fig. 1-12(B)

Plus	Minus
Water is a cheap and efficient heat transfer and storage medium.	High initial cost.
	Corrosion, scale, or freeze can occur causing damage or blockage.
Piping uses little floor space.	Leakage can cause considerable damage to the system and dwelling.
Circulation of water uses less energy for equal heat content.	Contamination of water supply is possible if poor design allows treated water in storage to enter domestic water system.
Adaptable to large buildings.	
Can provide energy for cooling.	Summer boiling can be problem.

A Heat Pump for Heating

A heat pump is correctly identified as a refrigerant machine that can extract useable heat from a source that is at a temperature too low for *direct* comfort applications.

It is important to recall that in the fundamentals of heat flow, it is a temperature *difference*—not temperature alone—that influences the amount of heat transferred.

For example: A gallon of water cooled from 55° to 35°F—a 20°F difference—gives off a greater quantity of heat than a gallon of water cooled from 210° to 200°F (10°F difference), although the latter is hotter to the touch.

Cold is only a relative term. Heat energy is almost always present. The exception is at absolute zero temperature (minus 460°F) where all thermal motion of atoms stops.

Thus, as long as the refrigerant circuit of a heat pump can produce a temperature *difference* (by providing a transfer medium colder than the temperature of the heat source), heat can be extracted. Remember, heat flows only from a higher to a lower temperature body.

For this reason, it is possible to use air, water and the earth itself as a source of heat—even though one normally thinks of these substances as lacking heat or being *cold* during the winter.

A solar collector can be used to heat air or water, so it is logical to combine a collector circuit with a heat pump. As the heat pump extracts heat from storage, the solar collector circuit can be used to replenish the supply. Figure 1-13 illustrates this concept schematically.

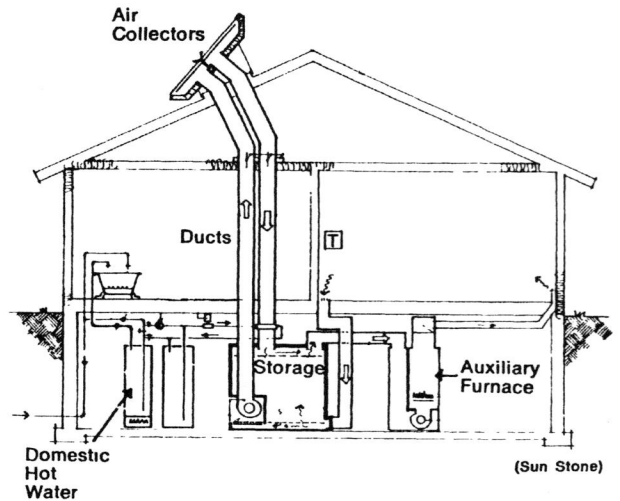

Fig. 1-12(A). All air solar heating system.

And, as was demonstrated at the University of Kentucky several years ago, it is possible (but not yet practical) to combine a solar collector with an earth source heat pump as well. In this concept, the earth is used as the storage "tank" and heat is added by the collector. Heat is then extracted by the heat pump via hundreds of feet of pipe buried in the soil.

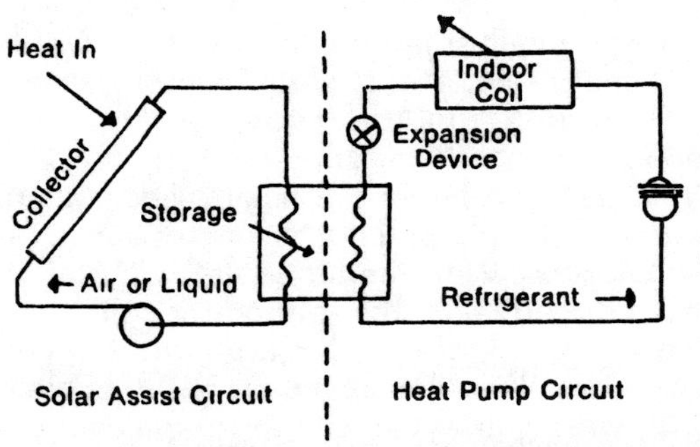

Fig. 1-13. Schematic of a Solar assisted heat pump.

Solar Cooling

Of the three general categories of space cooling methods for residential building—refrigeration, evaporative cooling, and radiative cooling—solar energy is only directly useful in refrigeration methods. Refrigeration systems cool by removing heat as it comes in contact with a cool refrigerating surface. In both the conventional vapor-compression systems using electric motors and heat pumps, and the absorption systems which use a fuel, solar heat could be used to drive the compressor in vapor-compression systems or replace the fuel in absorption systems.

Vapor-Compression Cooling

In the familiar vapor-compression cycle, an electric motor drives a compressor that "pumps" refrigerant gas from the relatively low pressure and temperature in the evaporator up to the pressure and temperature level existing in the condenser.

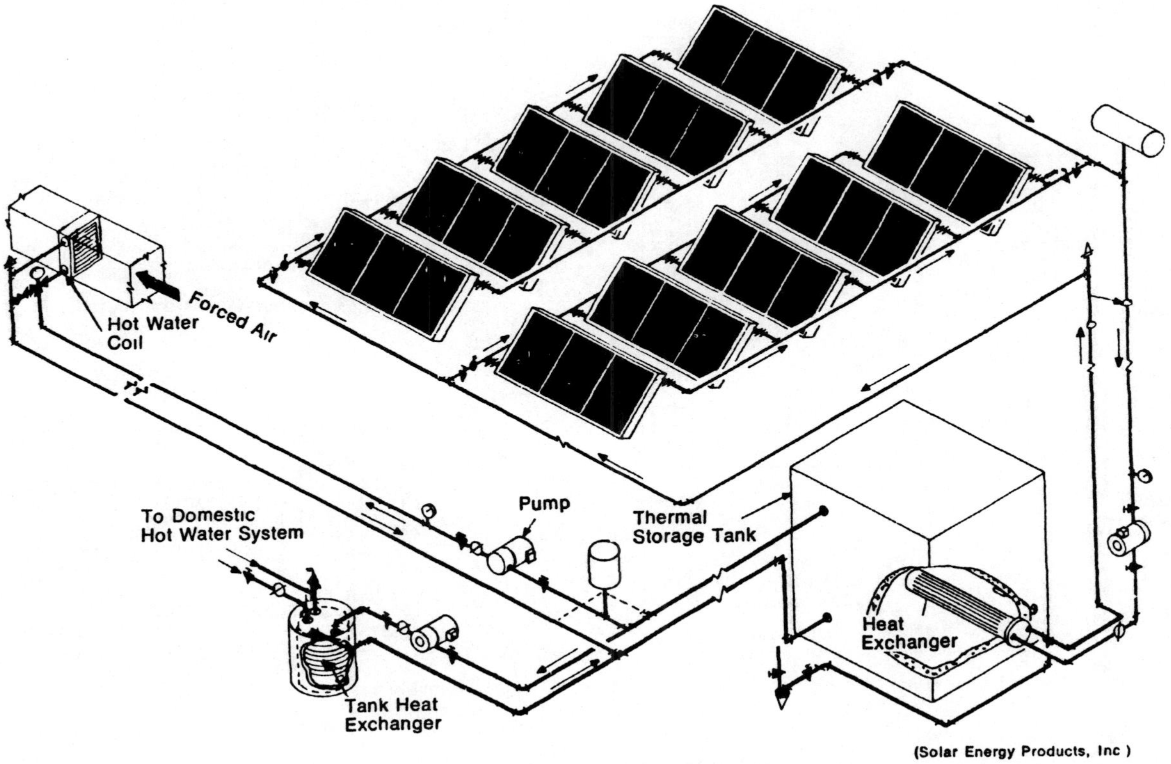

(Solar Energy Products, Inc)

Fig. 1-12(B). Hydronic solar heating system.

Absorption Cooling

In mechanical cooling, unwanted heat is absorbed at low pressure and temperature and then rejected at a higher, more convenient pressure and temperature.

In the absorption process, heat replaces horsepower. The compressor is replaced by an absorber-generator assembly in which a liquid is used to carry the refrigerant vapor from evaporator to condenser.

A closer look at the absorber-generator shows how refrigerant vapor from the evaporator enters the absorber where the vapor is absorbed by a liquid. The liquid solution is then pumped to the generator (which is at the same pressure as the condenser).

In the generator, the liquid is heated. This drives the refrigerant vapor out of solution and the gas continues on to the condenser. The liquid, now free of refrigerant, is returned to the absorber to repeat the process.

In conventional absorption cooling units, heat is supplied from gas or oil. In a solar assisted unit, a hot fluid provides the required heat.

Lithuim-Bromide Chillers. Of the absorption refrigeration currently in use, the Lithium-Bromide Chillers appear to hold the most promise. Water is the refrigerant, lithium-bromide is the absorbent. The system operates under vacuum and requires a cooling tower for water-cooled condensing. The system cannot use air-cooled condensing because of the low temperature in the absorber.

There are two types of chillers—air chillers and water chillers—which also describes their chilling function. With the air chiller, room air is circulated directly past the evaporator coils.

The water chiller requires a remote fan coil unit with room air cooled as it passes through the unit. The water chiller has an advantage over the air chiller in that it is possible to store chilled

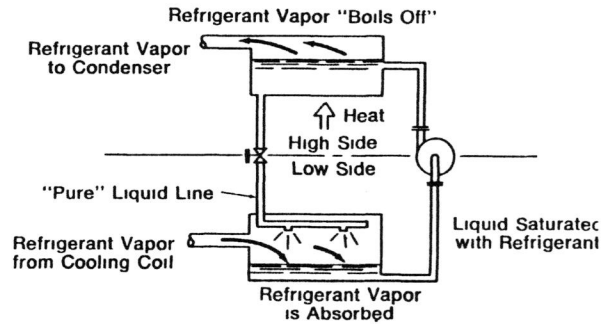

Fig. 1-14. Basic vapor compression refrigeration cycle.

Fig. 1-15. Generator-Absorber replaces mechanical compressor.

Fig. 1-16. Absorber-Generator details.

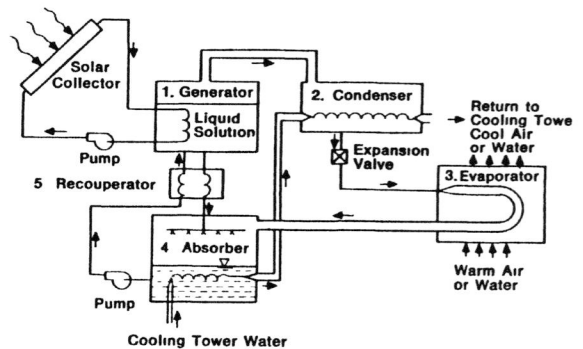

Fig. 1-17. Solor Powered Absorption Air Conditioner.

water when the building does not require cooling (as in the morning or at night). This enables the system to provide for a large peak cooling capacity. Typical requirements to provide 36,000 btu/h of cooling are a hot water supply of eleven (11) gallons per minute at 190°F. Cooling capacity obviously falls off drastically with lower temperature water supply from the collector circuit.

A Heat Pump for Cooling

A heat pump is another possible cooling unit. It is most easily remembered and described as an air conditioner that can pump out warm air when desirable, and then reverse the cycle when necessary and pump warm air in. It is a mechanical vapor compression system consisting of a compressor, condenser, expansion valve, and an evaporator. When in the cooling cycle, a heat pump is not a solar energy powered device. It is only in its heating mode that it can become a solar assisted device. This happens when solar heated transfer materials (air or water) are used as the heat supply to the heat pump.

Solar Rankine-Cycle Engine

Solar Rankine-Cycle Engine is a solar powered engine that can be used as an alternate source of power for the regular electric motor driven compressor in a vapor-compression refrigeration system. Solar heat can vaporize an organic fluid to drive a turbine which is coupled to a compressor. Thus, through the use of a Solar Rankine-Cycle Engine, a device such as a heat pump could become a solar related device in both cooling and heating modes. (See Figure 1-18).

Evaporative Cooling Plus Storage

Conventional evaporative cooling in arid and semi-arid parts of North America is accomplished by circulating outdoor air through a wetted pad. This simultaneously lowers the dry bulb temperature of the air and raises the wet bulb temperature. This cool but humid air is then circulated through the house for comfort cooling.

Strictly speaking, evaporative cooling cannot be "helped" or powered by a solar system. However, the rock (pebble) bed storage associated with an all-air solar heating system can be used along with an evaporative cooler to store cold at night for use during the day.

Here is how it works. At night, outdoor air is circulated through the evaporative cooler and the air is then circulated through the rock storage to subcool the rocks. This moist air is then exhausted to the out of doors.

In the daytime, hot room air is routed through the cool pebble-bed, sensible heat is removed and circulated throughout the house. Note that no latent heat (moisture) is removed in this process, thus, this technique is not well suited for all areas.

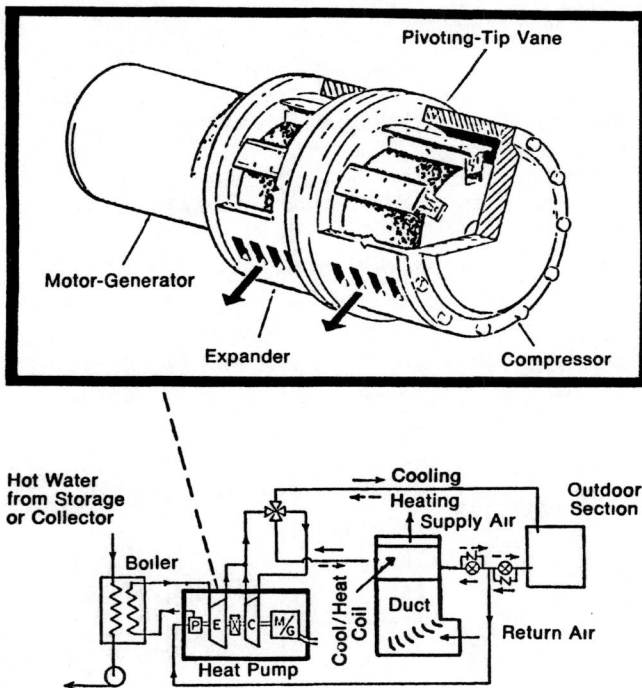

Fig. 1-18. Components of the Battelle Solar Heat Pump.

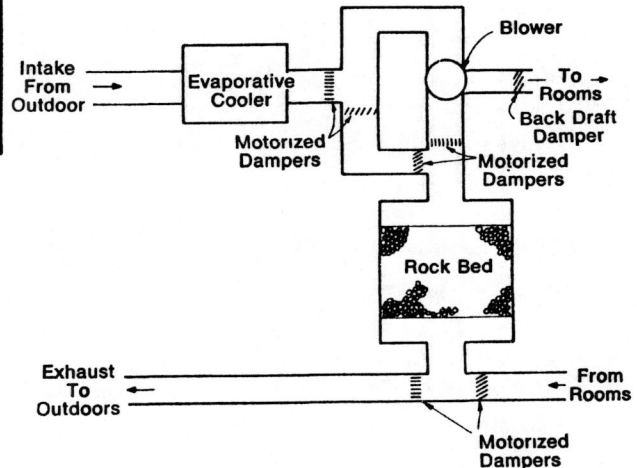

Fig. 1-19. Evaporative cooling with rock storage.

Other Cooling Possibilities

Besides providing comfort cooling via mechanical devices, it is also possible to *cool* using chemical reaction devices similar in operation to the widely used chemical dehumidifiers—liquid absorption and solid absorbent equipment.

In these devices, a desiccant—moisture capturing chemical—is used to dehumidify air and then solar heat is used to provide some of the energy to regenerate the desiccant after it becomes saturated with moisture. Evaporative cooling can also be added.

One innovative device under study is the Munter Environmental Control (MEC) unit shown in some detail in Figure 1-20.

Reverse radiation from the absorber surface of a flat-plate collector to the cold night sky can cool the absorber surface. It can also cool the water or air circulating through a collector. To use this principle, a shallow water pond covers the roof with sectionalized, retracting, insulated covers over the pond. At night, the covers are retracted to cool the water in the pond by both evaporation and radiation. They are then closed during the day to prevent the water from being heated. The cool water is circulated through fan-coil units to cool the rooms below. In the winter, the process is reversed to capture the sun's warmth during the day. The covers are closed at night to retain the heat which is radiated to the rooms below.

Evaporation of the water, which occurs naturally, increases its salinity and thus requires periodic draining. Possible freezing is a problem.

The generalized schematic (Figure 1-21) provides an idea for this type of system.

ECONOMICS

The exploding demand for energy, both in the United States and throughout the rest of the world, will continue to force the cost of energy up and will result in recurring energy shortages. Compared with other energy alternatives, solar energy for space heating and domestic water heating requires relatively high capital, although this factor is being offset by several factors: (1) increasing cost of other fuels, (2) intensive research to develop more cost effective systems, and (3) the potential for an excellent life payback cycle. A prototype system using circulating air has worked well since 1957 with no major problems and essentially no maintenance costs. Solar assisted heating is becoming economically competitive with heating by fuel oil, propane, and electricity in many parts of the country.

With present day equipment and prices, most solar systems have a fairly long payback. The energy is free but the equipment investment is relatively high. As additional square feet of collectors and larger storage tanks are added, one must look at the new investment versus savings. A solar system large enough to provide 100 percent of the heating requirement has a much longer payback period than one which does just part of the job. At present, most solar installations are sized to handle 50-70 percent of the total load. A backup conventional system sized to handle 100 percent of the load is normally included to handle conditions when no solar radiation is available and the stored heat has been depleted.

Location Differences

The economics of solar heating in different

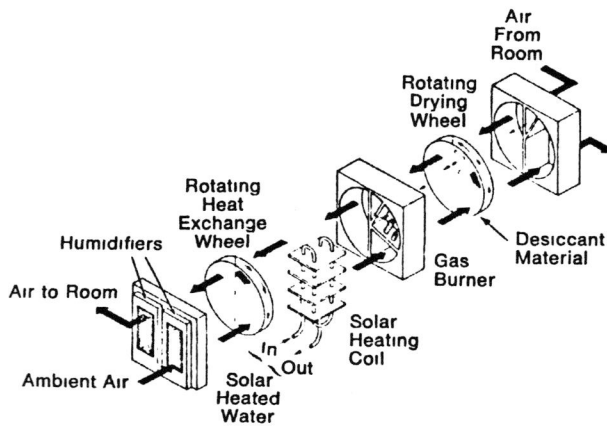

Fig. 1-20. Schematic of Adsorption Cycle (courtesy of the Institute of Gas Technology.)

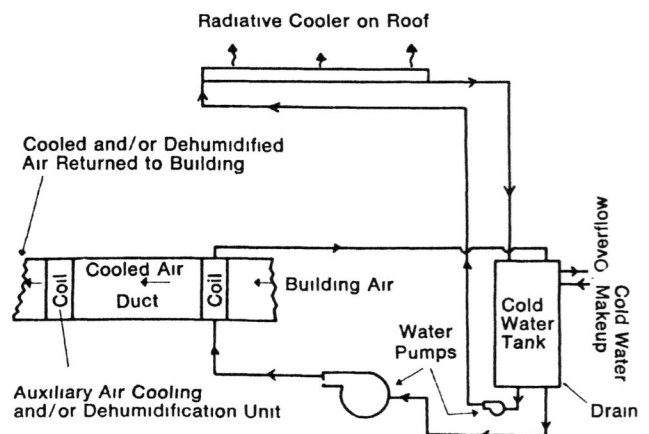

Fig. 1-21. Radiative Cooling System.

locations depends on a combination of the following factors:

1. Cost of equipment.

2. Amount of solar radiation available during the heating season.

3. Heating load of the proposed structure.

4. Cost of conventional heating energy.

5. The extent to which the solar heating system can be utilized outside the heating season, for example, domestic water heating.

Because of colder weather and greater heating requirements found in northern latitudes and excessively cloudy weather in coastal areas, it is difficult to predict whether or not solar heating systems will offset the costs of heating with other fuels until all the cost and weather factors are determined. For example, the greatest average amount of solar radiation in the United States is available in the Rocky Mountains—Southwestern area because of relatively clear weather conditions. Other areas of the country are not excluded, however, since experience indicates that solar heating is economically and technically feasible in many parts of the country at this time and will become more wide spread as conventional energy prices rise. Consider the following predictions based on the cost of energy in Fort Collins, Colorado in Table 1-1.

In a more general tone, the U.S. Department of Labor, Bureau of Labor Statistics presents the overview in Table 1-2. The only figures not supplied by USDOL are the projected 1980 utility costs. Estimates are based on distillation of data from industry, Edison Electric Institute, Arthur D. Little study, Energy Law (PL 94-163), and assuming natural gas wellhead deregulation before 1980.

Table 1-1

Fuel	Unit	Btu/Unit	Efficiency of Heating (%)	Cost* Per Unit ($)	Cost** Per Million Btu ($)	Predicted Annual Cost Increase (%)
Natural Gas	1,000 cu. ft.	880,000	80	1.02	1.45	17
Coal	2,000 lbs.	21,000,000	65	42.00	3.08	
Fuel Oil	Gallon	135,000	75	.38	3.75	15
Propane	Gallon	91,000	80	.35	4.77	20
Electricity	KWH	3,412	99	.02	6.10	7

* Currently the cost of electricity per KWH is 2c in Fort Collins, Colorado; 3c in Columbus, Ohio; and 9c in New York City.

** The costs in this column take into consideration the fuel's efficiency of heating.

Table 1-2
U.S. Utility Rate Comparisons
(Residential averages)

	Percentage of Increase Dec. 1967-Dec. 1975	Cost of Million Btu's 1973 vs 1975		Present Utility Costs	Projected 1980 Utility Costs Dollars	%
Gas	82.1%	$ 1.28	$ 2.33	$15.12 per 100 therms	$45.36 per 100 therms	300%
Oil	130.6%	$ 2.27	$ 5.24	$39c gallon (#2)	$54.6 gallon (#2)	40%
Electric	57.9%	$ 7.54	$11.91	4.5c KWH	6 3c KWH	28%

Couple the above information with a prediction that all these fuels (except electricity), will be unavailable in parts of the United States at any price for space heating, and electric space heating will be far more costly; and solar space heating becomes very realistic and competitive.

Proper System Sizing Saves Dollars

When a solar heating system is designed (sized) for a building, its basic cost can be minimized by first reducing the heating load of the building to a practical minimum. For example, a particular model of a collector may cost $14.00 per square foot. If 500 square feet were initially required for a specific house, the investment would be $7000 for collectors. If an additional $500 worth of insulation would reduce the required number of collectors to total 350 square feet, the savings in

collector cost would be $7000 minus (350 × $14.00) or $2100. Deducting the $500 additional cost for the insulation, the owner's net saving in first cost will still be $1600. Thus, the maximum energy cost savings per unit of initial investment can be realized. Also, unless dictated by unusual factors, it is not practical to size collectors to carry 100% of the load, therefore 50-70 percent of the annual heating requirements is a practical and economical range for the solar heating portion.

Further items which might help influence an individual to incorporate solar heating into the construction of a home would include the following possible cost savings:

1. Incorporation of systems into building design.

Table 1-3

City and State	Latitude Degrees °N	Elevation	Average January Temperature Degrees °F	January Solar Btu/(ft.² day)
Boston, Massachusetts	42	29	30.2	511
Cleveland, Ohio	41	805	28.4	456
Ithaca, York	42	950	23.	449
Lemont, Illinois	42	595	24.8	629
Lincoln, Nebraska	41	1,189	24.8	699
Medford, Oregon	42	1,329	37.4	434
Newport, Rhode Island	41	60	28.4	570
New York, New York	41	52	32	537
Salt Lake City, Utah	41	4,227	28.4	648
Sayville, New York	41	20	35	603
State College, Pennsylvania	41	1,175	28.4	511
Upton, New York	41	75	35	583

2. Ease of maintenance and repair. Warranties, guarantees, reliable product service, and replacement parts must grow at a rate to support installed systems.

3. First cost versus long term. Consumer education is essential to show the merits of life cycle costing (predicting how long the equipment can be expected to operate before replacement and what the real dollar amount would be) together with lenders willing to capitalize the solar equipment on long term financing. This kind of information is especially important to commercial building owners contemplating the design and construction (or remodeling) of large structures.

4. Safety, realiability, and efficiency. Solar heating and cooling hardware should be proven to be at least as safe, reliable, and efficient as conventional equipment.

Additional Costs Beyond the System

Some of the factors which must be considered when costing a solar heating system are:

1. Cost of solar assisted heating/cooling equipment.

2. Cost of auxiliary energy.

3. Cost of installation.

4. Cost added to building construction to accomodate the system.

5. Property taxation rates.

6. Loan interest rate.

7. Cost of maintenance and service including replacement parts.

8. Annual cost of insurance.

9. Credit for income tax savings.

Based on the above factors, it is essential that the solar heating/solar cooling system provide the maximum possible return for dollar invested. These returns are based on savings over the installation, operating, and servicing of conventional heating systems.

Climatological Considerations

Further consideration must be given to the climatological characteristics in the area under consideration. The chart in Table 1-3 is provided to emphasize that assumptions about climatological characteristics are not always correct.

By simply comparing cities which are found in a 70 mile wide belt that stretches from coast to coast, a difference of 12°F is found in the average temperature for the month of January (Ithica, New York's 23°F versus Medford, Oregon's 37.4°F). At the same time and in this same belt, January Solar Btu's varied by 265/day. Medford, Oregon is low with only 434 Btu's (although Ithaca, New York is only 15 Btu's better at 449) while the high side is held by Lincoln, Nebraska with 699 Btu's (although its average daily temperature during the period—24.8°F—is only 1.8°F higher than Ithaca, New York's 23°F). The point of this exercise is that many factors must be carefully analyzed when climatological characteristics are considered. Boston, Massachusetts, receives 5" more precipitation per year than the usually wet city of Seattle, Washington. Flagstaff, Arizona, receives an average of 96" of snow per year which is more than Burlington, Vermont's 80" average. Since these factors contribute to the cost of the system, it becomes increasingly important to know the geographical area you serve.

SUMMARY

A basic overview concerning essential solar heating/solar cooling components and the economic considerations which influence a person contemplating a solar energy system has been presented. One need not spend much time attempting to determine whether the utilization of solar energy for residential and commercial heating is necessary or feasible. The lack of fossil fuels makes it imperative to design efficient heating systems which utilize solar energy to provide for our heating and cooling comfort.

SOLAR RADIATION

There are six basic forms of energy useful to man—chemical, radiant (solar), heat, nuclear, mechanical, and electrical energy. Heat is considered a *lower* form of energy since all other forms can be converted to heat fairly easily, but the heat energy cannot be readily converted to the other forms.

The sun provides a highly concentrated source of radiant energy which can be converted to heat. Consider that the sun has projected radiant energy across 93 million miles to earth, and that earth offers only a very small target to the sun's outgoing energy. The tremendous total amount of energy which the sun must be giving off can be realized. Even though the earth's share of this radiated energy is relatively small, a great number of the earth's inhabitants are more than casually interested in the portion that our planet manages to intercept.

ENERGY AND SOLAR RADIATION AND LIFE

Life exists because of the sun's energy. Sunlight that was received in past ages is also the reason for the existence of our fossil fuels—coal, oil, and natural gas—which are now becoming in short supply. Sunlight is responsible for the existence of living forms of fuels, such as celluloses and starches, and their by-product, alcohol. In addition to energy, the sun is responsible for earth's existing climate. As small as the interception of solar energy by the earth may be, it is sufficient to heat the atmosphere and keep the earth from turning into a solid, frozen sphere.

THE NATURE OF SOLAR RADIATION

What is the nature of the radiation received from the sun in such vast quantities? For this course, the sun is a sphere, the outer skin temperature of which is approximately 10,000°F and from which is emitted energy in the form of radiation. Figure 2-1 shows the electromagnetic spectrum of the sun's rays and depicts seven different types of radiation, although our interest only involves the middle three: ultraviolet, visible, and infrared rays. This is not to imply that there are no effects from gamma rays, X-rays, or radio waves, but these rays are not important at present to the heating and air conditioning industry.

THE COMPOSITION OF SOLAR RAYS

The radiant energy waves can easily be arranged in increasing order of wave length as illustrated in Figure 2-2. This illustration also provides

an idea of the heat generated by these different waves, evident in the vertical scale on the left-hand side which is marked in energy intensity in British thermal units per hour per square foot [Btu/(ft^2 h)]. This scale represents the amount of heat that would result if waves of a given wave length were completely absorbed by a flat surface one square foot in area for one hour. The higher the curve, the greater the heat. The instrument used to measure solar energy is called a *pyrheliometer*. It consists of a very sensitive black disc which when exposed to the sun's rays, is heated. The black disc is connected to one end of a thermocouple. As the disc heats, a voltage is produced at

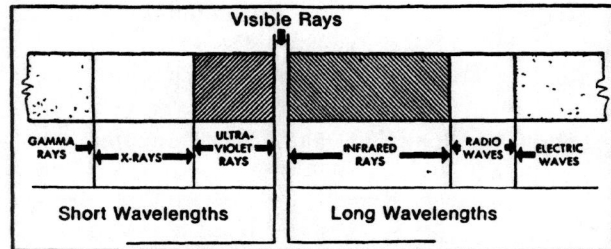

Fig. 2-1. The Electro-Magnetic Spectrum. Radiation from the sun consists primarily of ultraviolet, visible, and infrared radiation.

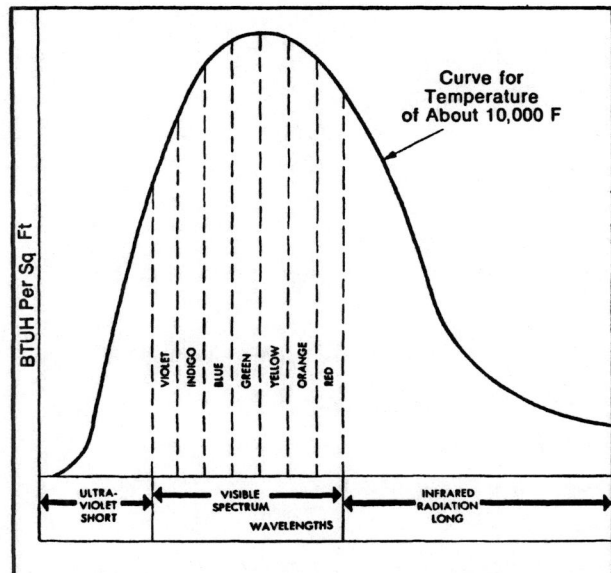

Fig. 2-2. Radiant energy from the sun.

the other end of the thermocouple (just like a thermocouple heated by a pilot light). Since the voltage output will vary with the temperature of the disc, the output can be calibrated to indicate Btu's per hour per square foot of radiation falling (See Figure 2-3).

DIFFERENT UNITS OF MEASURE

The above description is an accepted manner of expressing a unit of solar energy—$Btu/(ft^2 \cdot h)$. However, other units may be found when reading solar data. Now is an ideal time to learn to recognize these units and to be able to convert from one unit to another. The units are generally either *Energy Density* expressions or *Power* expressions and are usually expressed as follows:

ABBREVIATION	UNIT
Energy Density	
Btu/ft^2	British Thermal Units per square foot
KJ/m^2	Kilojoules per square meter
Langley (cal/cm^2)	Calories per square centimeter
Power	
$Btu/(ft^2 \cdot h)$	British Thermal Units per square foot per hour
$KJ/(m^2\,h)$	Kilojoules per square meter per hour
Langley/min	Calories per square centimeter per minute
W/m^2	Watts per square meter

To convert these units from one set to another, use the following conversion information.

To Convert Into Btu/ft^2		To Convert Into $Btu/(ft^2 \cdot h)$	
Multiply	**By**	**Multiply**	**By**
Langleys	3.69	Langleys/min	221
KJ/m^2	088	$KJ/m^2\,h$	088
		W/m^2	316

The following example will show the use of these data. Assume that the annual average daily solar radiation for Blackstone, Massachusetts, is 352 Langleys per day. To convert this to Btu/ft^2, multiply by the conversion factor for Langleys to Btu/ft which is 3.69:

$$\begin{array}{ccccc} 352 & x & 3.69 & = & 1299 \\ \text{Langley's} & & \text{Conversion} & & Btu/(ft^2\,d) \\ \text{day} & & \text{Factor} & & \end{array}$$

SOLAR INTENSITY

The intensity of the sun's energy output is quite constant. Just outside the earth's atmosphere, this intensity has been determined to be 1.940 Langleys/min (1,940 x 221 = 428.7 $Btu/(ft^2\,h)$) at the average earth-sun distance and is called the "solar constant."

The Solar Spectrum

As mentioned earlier, the radiation from the sun can be separated into three major energy regions:

1. The short-wave-length radiation at the left of Figure 2-2 which is called *ultraviolet* or UV radiation and is not visible to the human eye. These are the rays that produce our suntan or sunburn. A considerable amount of the UV radiation which enters the outer atmosphere is absorbed there and does not reach the earth's surface.

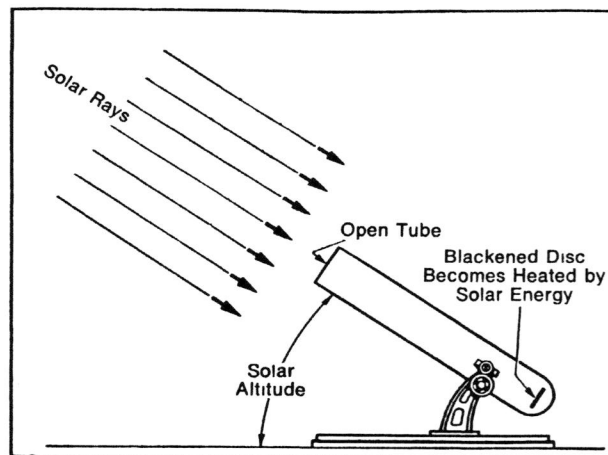

Fig. 2-3. The instrument used to measure solar energy is called a pyrheliometer.

2-2

2. The middle wavelengths in Figure 2-2 are referred to as the *visible spectrum*, since these are the wave lengths which can be seen by the human eye. When this white light is passed through a specially constructed, traingular piece of glass called a prism, the varying speeds at which these wavelengths travel cause it to be broken into colors that vary from violet through the blues and greens to the reds. As indicated by Figure 2-1, the visible spectrum is an extremely narrow band when compared to the ultraviolet and the infrared bands.

3. The long-wave-length is referred to as *infrared radiation* (IR), or heat radiation, although the latter term is not a good one since both ultraviolet and infrared will heat an object. Practically any object can be considered to be radiating infrared, even a cake of ice. The amount of infrared radiated will depend on the item's temperature. Materials with a temperature below 800°F. emit *only* infrared, with no visible or ultraviolet radiation. Thus, the radiation emitted from surface temperatures below 800°F is quite different in quality from the radiation of the sun.

A solar collector absorbs radiation from all three of these regions. The ultraviolet and visible, as well as infrared, are converted into heat on the surface of the collector.

Energy Reaching the Earth
The energy reaching the earth ranges from 75 to zero percent of the "outer space" value of 100 percent. From outer space, the energy hits the upper atmosphere and some is reflected back. Some which passes through is reflected by the tops of clouds and by dust. The ozone layer in the upper atmosphere absorbs much of the ultraviolet radiation, while carbon dioxide, oxygen, and water vapor also absorb radiation. Dust and clouds scatter some of the radiation as illustrated in Figure 2-4.

Note that radiation is termed *direct* radiation if it has *not* been scattered, *diffuse* if it has. On clear days, diffuse radiation may represent 10 percent (direct equalling the other 90 percent) of that day's radiation. On cloudy days, diffuse radiation may represent all of the solar energy available for use.

How Much Solar Radiation Actually Reaches Us?
How much radiation actually does reach the earth? Assume that a house is 40 feet long by 30 feet wide and occupies a ground area of 1200 square feet (30 x 40 = 1200). Ignoring such factors as shadows, the sun will shine on a minimum flat area of 1200 square feet. On a clear day, at noon, the solar energy received by 1 square foot of flat surface area is about 290 Btu/h. Thus, for the 1200 square feet of surface, the total heat energy absorbed is 348,000 Btu/h (1200 x 290 = 348,000).

How much heat does this represent? In practical terms, this amount of heat will melt 29 tons of ice in one day (12,000 Btu/h equals the cooling capacity of one ton of ice. Divide 348,000 by 12,000 and the answer is 29). This is how an air conditioner gets its ton rating. Fortunately, houses are not greenhouses and are designed to reflect away most of this heat, so little solar energy gets inside of the houses.

Monthly Variations
Solar energy radiated to a horizontal surface at any location on earth, if averaged over a month, shows a month to month variation. This fact is due to both seasonal changes in weather and the changing angular relationships between the sun and the earth's surface. In the winter, the sun is lower in the sky than in the summer. The resultant angle between the sun and the horizontal surface reduces the portion of radiation intercepted by the surface. Figure 2-5 illustrates how the angle of the sun to earth differs in summer and winter. This aspect can be seen in more practical terms in Figure 2-6: a roof overhang is used to reduce the amount of solar radiation entering a house during the summer.

This variation in the sun's position in our sky occurs because, as the earth follows its orbit, the tilt of the earth's axis changes our relationship to the sun. Generally speaking, the angle of the sun

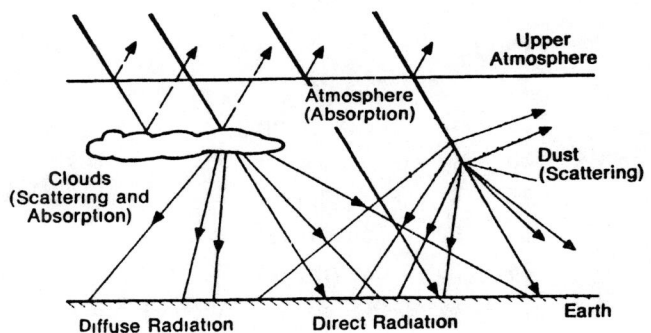

Fig. 2-4. Atmospheric Effects on Solar Radiation Reaching the Earth's Surface.

above the horizon at noon is about 25° in winter, and 75° in summer for latitudes near 40°N as shown in Figure 2-5. Since this angle differs as the latitude differs, monthly averages vary in each locality. Table 2-1 is included to show monthly variations on a horizontal surface for selected cities in the United States. Note, for example, that Chicago averages considerably *less* solar radiation than Tucson or Miami in December. Solar data for a number of cities are listed in Appendix A.

Hourly Variations

Another item for consideration is the hourly position of the sun, which is clearly illustrated in Figure 2-7. Summer days have more hours of sunlight and the area exposed to this solar radiation is greater.

Early morning sun is at a very low angle and the solar rays must pass through a larger thickness of atmosphere than at noon time (see Figure 2-8.) This is a prime reason why noon sun is "stronger." If the sun was tracked with a pryheliometer through the day, the Btu's received would vary in a manner shown in Figure 2-9.

If the solar energy received each hour was plotted on a fixed horizontal surface that same day, the pattern would be indicated by the solid line in Figure 2-10.

As expected, the greatest intensity still occurs during the noon hour, but the energy received by the horizontal surface was less than in the pre-

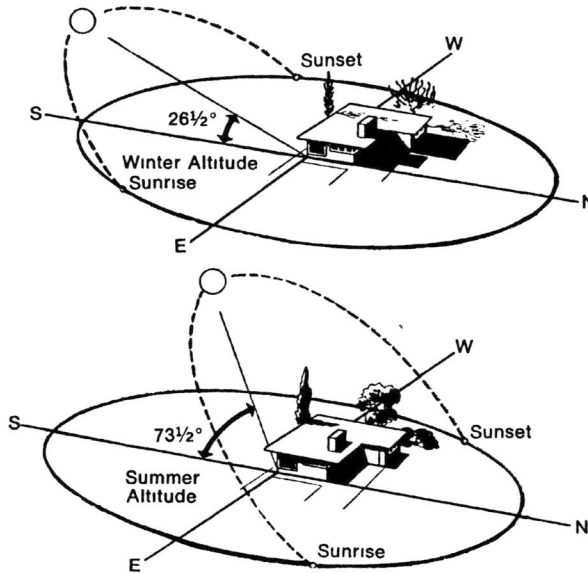

Fig. 2-5. SUN'S position in sky affects amount of heat absorbed. Winter sun is hotter but lower in the sky than summer sun. Angles shown are for 40 deg N Latitude (Chicago).

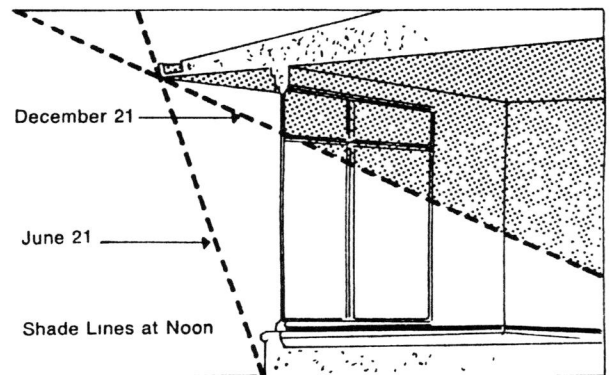

Fig. 2-6. High summer sun is block out by roof overhang while in winter, lower sun permits rays to penetrate house.

Latitude						
°	'	CITY	DECEMBER	MARCH	JUNE	SEPTEMBER
41	52	Chicago, Illinois	280	836	1688	1153
32	13	Tucson, Arizona	1093	2010	2610	2139
38	54	Washington, D.C.	541	1178	2054	1351
25	47	Miami, Florida	1174	1808	1955	1646
64	50	Fairbanks, Alaska	22	858	1940	677
34	30	Los Angeles, California	905	1690	2272	1855

Table 2-1

Monthly Variations in Energy on a Horizontal Surface in Selected United States Cities (Btu/(ft^{2}d)

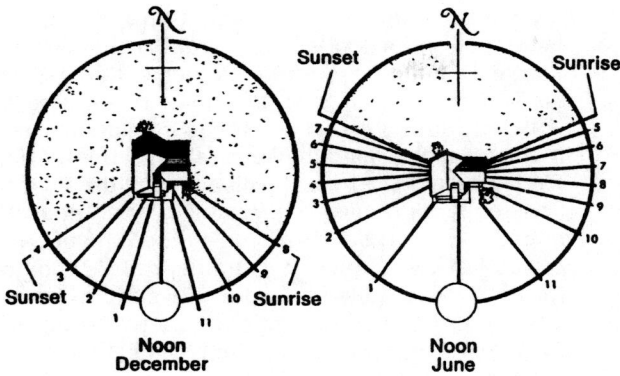

Fig. 2-7. With the sun higher in the sky in summer there are more daylight hours in summer than in winter. Note time of winter and summer sunrise and sunset.

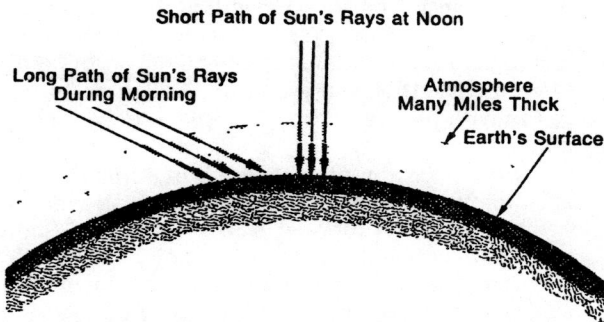

Fig. 2-8. Interception of sun's rays by earth's atmosphere.

vious case when the sun was tracked (dotted line is from Figure 2-9). The curve indicates the readings were taken on a clear day since it is smooth. The presence of clouds would have caused breaks in the curve.

Vertical Surfaces

While roofs are generally considered to be horizontal or tilted surfaces, building walls are not. However, walls are able to absorb solar radiation just like horizontal surfaces. The only difference is that their exposure time to the sun's rays are different. (See Figure 2-11.) Another item of interest in Figure 2-11 is that the north wall is not included. But, since the north walls of buildings in the northern hemisphere are not exposed to direct solar radiation in winter, a reading would prove useless since it would be a flat line zero. North walls can receive some diffuse radiation, however; thus, the surface direction (north, south, east, west) and surface tilt (horizontal, vertical, etc.) all affect the amount of solar radiation actually intercepted.

Collector Tilt

It is advantageous to tilt the solar collector so that it is perpendicular to the sun's rays as illustrated with the pyrheliometer (Figure 2-3). Figure 2-12 further illustrates this advantage by showing the increase in energy intercepted when a collector is tilted from the horizontal. The optimum tilt occurs when the angle of the collector is the same as the incoming radiation. The maximum energy would be intercepted if the collector

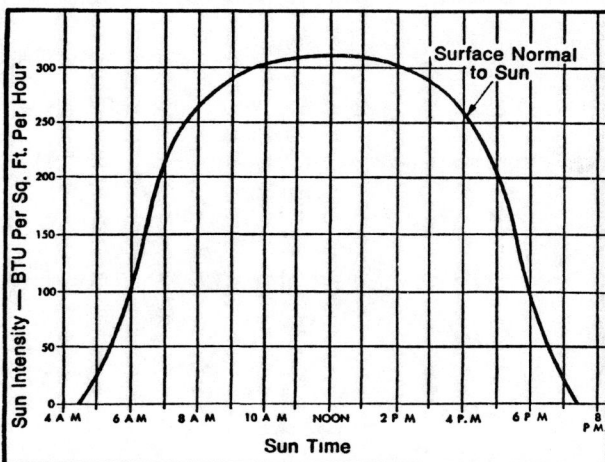

Fig. 2-9. Energy received by a surface kept normal to sun in summer.

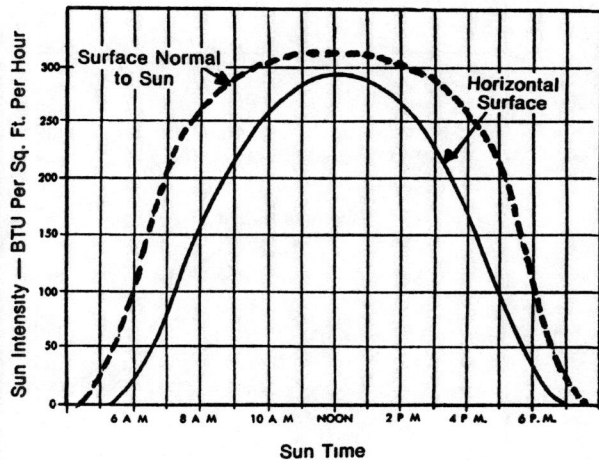

Fig. 2-10. Energy received by horizontal surface is less than that received by a surface kept normal to a summer sun.

2-5

were to track the sun across the sky, but tracking collectors are very costly and bulky for home installation.

The rule generally followed for the tilt of the collector in the northern hemisphere is to face the collector to the south. The angle of the tilt is latitude plus 15 degrees for heating or minus 15° degrees for cooling. For example, Woonsocket, Rhode Rhode Island, is located directly on latitude 42°N. If a collector were positioned for heating, its angle would be 57° (42° + 15° = 57°). If positioned for cooling, the collector's angle would be 27° (42° - 15° = 27°). When the collector is to be used for both heating and cooling, a reasonable rule is to have the angle of the collector equal the latitude. Thus, for Woonsocket, Rhode Island, the angle of the collector used for both heating and cooling would be 42°.

Collector Orientation

Since maximum solar intensity occurs at

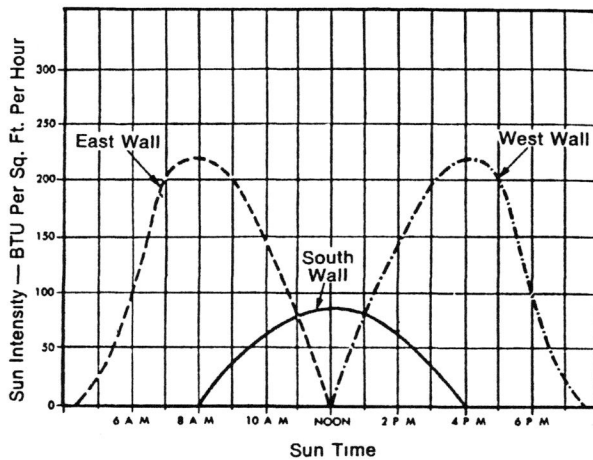

Fig. 2-11. Solar radiation falling on a building's walls in summer. Note times of maximum intensity.

Fig. 2-12. Effect of tilting the collector on energy intercepted.

noon when the sun is due south (in the northern hemisphere), a collector should face directly south. If building conditions make this impossible, a variation of ± 15 degrees can be tolerated without serious effect on the solar radiation collected. Keep in mind, however, that an orientation 15 degrees east of south will advance the time of peak collection one hour. A similar orientation 15 degrees west of south will delay the time of peak collection one hour. For example, if the collector location is partially shaded in the early morning, aiming the collector west of south would decrease the morning collection while increasing the better afternoon collection.

Collector Potential

Table 2-2 provides one manufacturer's example of maximum heat output from a single 27 square foot liquid "cooled" flat plate collector for winter months in selected cities. Values are in thousands of Btu's (per month). Utilized heat would be somewhat less depending on building load and storage losses. (Note: Collector output varies with different designs. This is only one example.)

How does the collector output compare with a building's monthly needs? Consider a well insulated house of modest size located in Columbus, Ohio, with a design heat loss of 36,000 Btu/h. The home could require as much as 14 million Btu's in January for heating. From the following Table 2-2, this particular collector panel during January in Columbus could provide some 280,200 Btu's at best. Therefore, 50 of these collector panels would be needed to satisfy the total January Btu load on this house—an uneconomical solution.

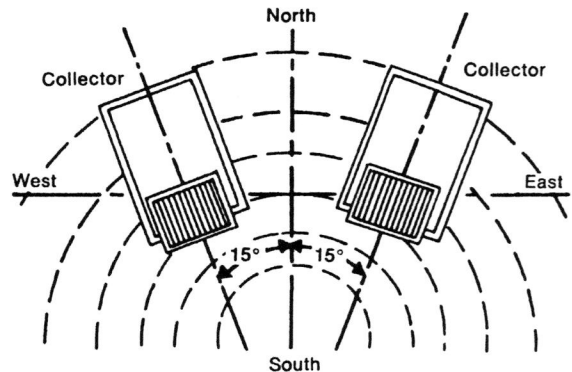

Fig. 2-13. Facing collectors other than due south has modest impact on energy collected. A 15 degree variation can be tolerated.

Practical Data

The amount of potentially useful solar energy reaching a surface through clouds, haze, and other atmospheric barriers varies from location to location, month to month, day to day, and hour to hour. Fortunately, heating technicians need not be astronomers and track the erratic movements of the sun. The heating technician need only be concerned with specific monthly averages in a given location to design practical systems by any number of suggested procedures.

SUMMARY

The first unit of this course has now been completed. The purposes were to develop an understanding of why solar energy is so important at this point in our history, what constitutes a solar heating and solar cooling system, and the basics of solar radiation. These understandings are necessary for progressing further in this course and when learning how to apply this knowledge in the design and installation of solar heating systems.

Table 2-2. Example of collector output, Btu/month/panel.

MONTHLY OUTPUT,/COLLECTOR PANEL — MBTU									
°Latt. North	Location	Collector Tilt	October	November	December	January	February	March	April
48.2	Glasgow, Montana	60°	496 2	338.8	264 3	333 6	436.7	536 7	450 5
43 6	Boise, Idaho	55°	543 2	387 4	288.3	313.1	404 4	494.4	481.9
40.0	Columbus, Ohio	55°	420 0	257.9	259 3	280.2	312.1	394 0	360.8
35.4	Oklahoma City, Oklahoma	50°	637 3	554.2	508 1	504.6	494.2	556 3	481 9
40	Salt Lake City, Utah	55°	536 6	407 4	392 2	345.4	406 6	475 3	440 4
29 5	San Antonio, Texas	40°	660 6	537.6	497 3	527.5	529.7	590.8	484 4
32.8	Fort Worth, Texas	45°	668 8	579 8	498 6	488 8	497.4	598 6	511.6
40 3	Grand Lake, Colorado	55°	509 9	386 4	370.4	390 8	450.9	410 5	415 0
42 4	Boston, Massachusetts	55°	425.4	307 6	279 1	303.0	319 2	389.7	339 8
27.9	Tampa, Florida	40°	660 0	656 6	610.0	646.7	608.0	670.0	578.0
33.4	Phoenix, Arizona	45°	777 0	663.0	589 7	606.2	655.7	756.0	678 0
33.7	Atlanta, Georgia	45°	566.3	489 5	422.7	423.8	441.7	518 0	499 0
35.1	Albuquerque, New Mexico	50°	719.2	628.7	580 8	604.9	592 9	682.3	588 0
40.8	State College, Pennsylvania	58°	468.6	307.5	255.8	280.6	312.9	405.8	375 5
42 8	Schenectady, New York	55°	381.8	242.1	315.1	281.7	317.9	365.2	319.5
43.1	Madison, Wisconsin	55°	465.5	306 7	293.9	321.0	343 5	442.6	370 2
33 9	Los Angeles, California	50°	604 3	577 2	540 4	535.4	556.8	633.0	482.9
45 6	St Cloud, Minnesota	60°	419 8	309.3	274 4	362 7	416.3	482.6	381.6
36.1	Greensboro, North Carolina	50°	537 8	465 1	396 0	409.5	430 2	481.4	456.4
36 1	Nashville, Tennessee	50°	556.0	424 0	354 5	325.7	380.7	456 8	438.2
39 0	Columbia, Missouri	50°	559.1	438.7	339 3	356 7	390 9	485 7	440 9
30 0	New Orleans, Louisiana	40°	589.7	506 8	390 5	415 8	396.6	473.5	448.2
32.5	Shreveport, Louisiana	45°	590.4	475 3	419.6	459.6	452.0	534.7	461.1
42.0	Ames, Iowa	55°	378.4	262 9	197.7	238.2	294.9	368.2	354.7
42.4	Medford, Oregon	55°	455.1	298 4	213.4	255.8	343.4	434.1	412.3
44 2	Rapid City, South Dakota	60°	550.9	436 3	383.3	405.1	453.2	518 0	420.3
38.6	Davis, California	50°	719.3	536.9	401.3	448.9	515.6	666.1	647.2
38.0	Lexington, Kentucky	50°	616 5	477.4	377.8	359.6	411.1	487.5	479.9
42.7	East Lansing, Michigan	55°	415 0	261.7	230.1	247.5	314.2	387.4	313 4
40.5	New York, New York	55°	505.8	389.4	328.0	357.1	396.6	451.9	381.8
41.7	Lemont, Illinois	55°	477.1	352.7	321.4	343.6	373.9	450.3	365.4
46.8	Bismark, North Dakota	60°	491.1	346.6	279.9	335.6	405 7	476 6	411.9
39.3	Ely, Nevada	55°	636.7	559.0	475.8	481.1	512.4	597.5	483.4
31.9	Midland, Texas	45°	651 5	596.2	540.7	543.0	543 1	648.9	562.1
34.7	Little Rock, Arkansas	50°	578.4	470.0	400.0	383.6	408.1	486.5	435.4
39.7	Indianapolis, Indiana	55°	501.7	354.3	293.2	300.1	332.1	417.6	360.4

(Fedders)

SOLAR COLLECTORS

The collector is the "furnace" of a solar heating system. This is the component which is exposed to the solar radiation from the sun and starts the process of harnessing solar energy into a reliable heat source for homes, business, and industry. Of all the elements in a solar assisted heating system, the collector is the component *least* familiar to the heating and air-conditioning technician. For this reason, it will be considered in some detail.

Several phenomena in nature compare to the collector principle. One example of solar radia-tion collecting is that of the almost unbearable heat that is felt when entering an automobile which has been parked in the sun with the windows closed. Solar energy has passed through the glass and has been absorbed by the fabric, metal, and/or plastic parts inside the car, then re-radiated as longwave energy which now cannot escape through the windows and is trapped as heat inside the car. As a result, dark-colored upholstry or other components like the dash or steering wheel may be extremely uncomfortable to touch.

The same scientific phenomenon related to the conversion of solar radiation into heat in your sun-baked auto occurs in a solar assisted heating system. A specially designed component, called a *solar collector,* efficiently converts trapped solar radiation into a useable heat source.

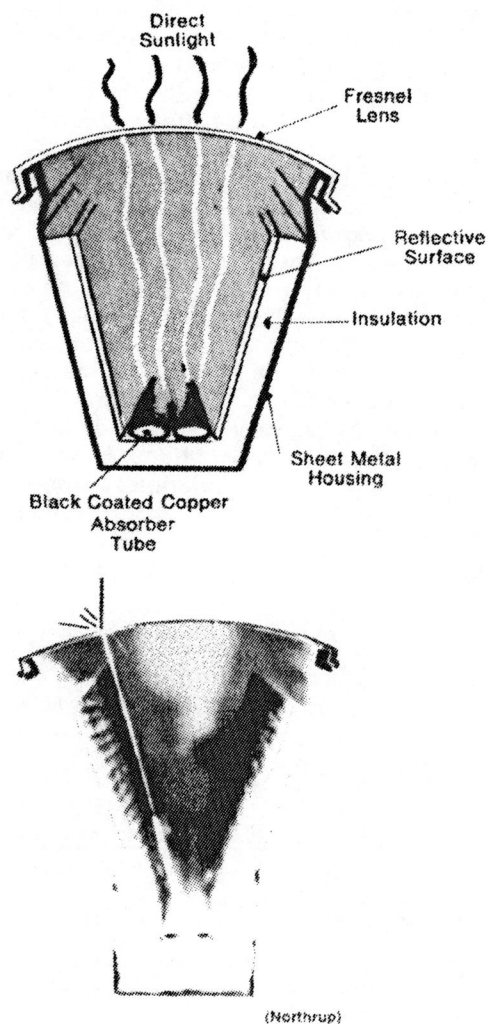

Fig. 3-1. Cross section of concentrating collector.

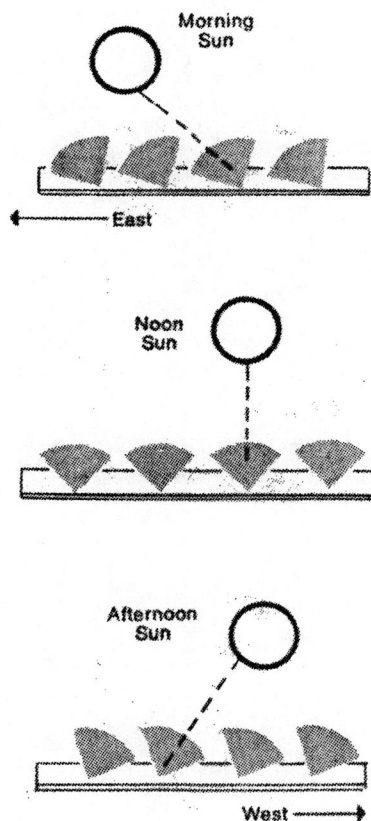

Fig. 3-2. Concentrating collector tracking sun.

TYPES OF SOLAR COLLECTORS

There are two basic types of solar collectors: the *concentrating* collector and the *flat plate* collector.

Concentrating collectors are shaped to "gather" *direct* solar radiation and produce *high* temperatures. They utilize the principle of focusing only the direct radiant energy by means of lenses or reflectors on a small absorber area to collect heat energy. In order to function properly, the concentrating collector must *track* the sun. This is done by mounting the device on a pivot so the collector can be moved to point toward the sun as it moves from the east to the west horizon during the day.

Flat plate collectors are termed *low* temperature collectors (Figure 3-3) and will function at various levels of efficiency depending on the time of day (position of sun) and ambient conditions (sunny to cloudy weather). They are permanently mounted in a southerly direction in the northern hemisphere at an angle to optimize collection and "absorb" both direct and diffuse radiation for a specific geographical location as discussed in Lesson Two.

Each collector configuration has its own relative effectiveness and efficiency. For this course, the study of collectors will be limited to the *flat plate solar collector*. Factors such as the state of the art and costs have made these devices more popular for comfort heating.

WHAT DOES THE FLAT PLATE SOLAR COLLECTOR DO?

A flat plate solar collector intercepts the solar energy from the sun and converts it into another form of energy—heat. This conversion is achieved by absorbing the sun's radiation into a thin black metal surface. The heat is conducted through the metal into a fluid medium (liquid or air) which transfers that fluid by a pump or fan to another part of the heating system.

To better understand how the collector functions, study Figure 3-4 and learn the terms to be used in the following discussion.

Solar radiation strikes the transparent cover of the collector. Some radiation is *reflected* and a tiny amount *absorbed,* but most is *transmitted* to additional covers where reflection, absorption, and transmission are repeated. Finally, the radiation reaches the absorber, which is usually made of metal. Here, the solar radiation is almost totally absorbed and heats the metal. The metal conducts the heat into the transfer system which circulates either a liquid through tubing or air through a duct to other sections of the solar heating system. The insulation behind the transfer system restricts heat loss through the back and sides of the frame. Excessive heat loss through the front of the frame is prevented by the

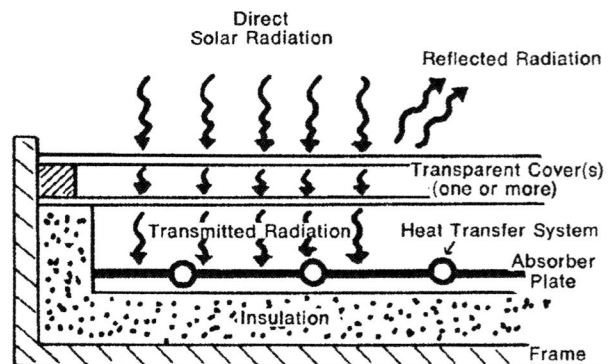

(Libby-Owens-Ford)

Fig. 3-3. Flat plate collector.

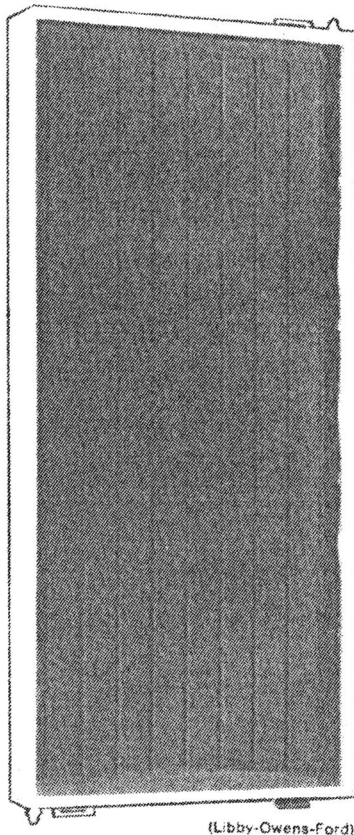

Fig. 3-4. Flat plate collector terminology.

cover(s) and air space(s) between covers much like the thermal effect of double glazed windows.

When the solar energy (composed of ultraviolet, visible, and infrared rays) passes through the air and glass and strikes the blackbody, the waves are absorbed. At this point, a phenomenon called *wavelength conversion* occurs. Some energy is reradiated from the absorber as longwaves and strikes the inner surface of the transparent cover. The composition of the glass cover causes it to be opaque to most longwaves and the heat is trapped and therefore is not lost. (This also explains the heated interior of a closed automobile parked in the sun). This so-called "greenhouse" effect can be explained by referring to Figure 3-5.

A relationship between radiant energy distribution (Btu/ft²) and wavelength was shown previously in Figure 2-2 in Lesson Two. Now, in addition to the curve for the sun at 10,000°F, the energy distribution for a fireplace (approximately 600°F) and a collector surface (assumed to be at 200°F) has been added.

Note that, as the temperature of the radiating body decreases, the curves shift toward long wavelengths. A fireplace gives off some radiation in the visible range, but a surface at 200°F emits only invisible infrared radiation which cannot be seen by the naked eye. In fact, any surface at 1000°F or less radiates entirely in the infrared range. Here again, the glass is essentially opaque to infrared or longwave radiation, and the longwaves are absorbed by glass. Then, some are reradiated back toward the absorber rather than allowed to pass through the glass.

WHAT ARE FLAT PLATE SOLAR COLLECTORS MADE OF?

To further develop the understanding of solar collectors, a more detailed explanation of the materials used to manufacture collectors are now presented. Keep in mind that the technology of solar collectors, as well as the whole field of solar heating, is advancing. As a result, solar heating components and systems will continue to undergo change at a fairly rapid rate based on research and the development of products and processes.

Transparent Covering

One or more transparent covers may be used on a collector. The optimum number of covers is determined by the collector design, collector use (heating/cooling comfort), and the annual, average outdoor temperature for a specific geographic area. No covers may be needed where average annual temperature readings are above 70°F because relatively little deviation will occur from comfortable temperatures. Two covers are required where the temperature averages 50°F or below for the year. While each cover reduces frontal heat loss from the collector, each additional cover also reduces the amount of solar energy transmitted, to some extent, so compromises in collector design must be made.

Materials Used. From outward appearances, it would seem that both *plastic* and *glass* would be acceptable materials for collector cover construction. They have many of the same characteristics of light and heat transmission. However, interior collector temperatures can easily reach 300°F; this would cause deformation of some types of plastic materials.

Several methods are being researched to increase "solar" efficiency of glass. One method is to reduce the solar energy reflected from the surface of the top glass. Another concept is to use a low iron-content glass that will absorb less energy. A third approach would be to coat the lower surface of the plate nearest the absorber with a substance that would help redirect more longwave radiation back toward the absorber plate.

Cover glass must be selected that will be stable in the outside environment. It needs to be

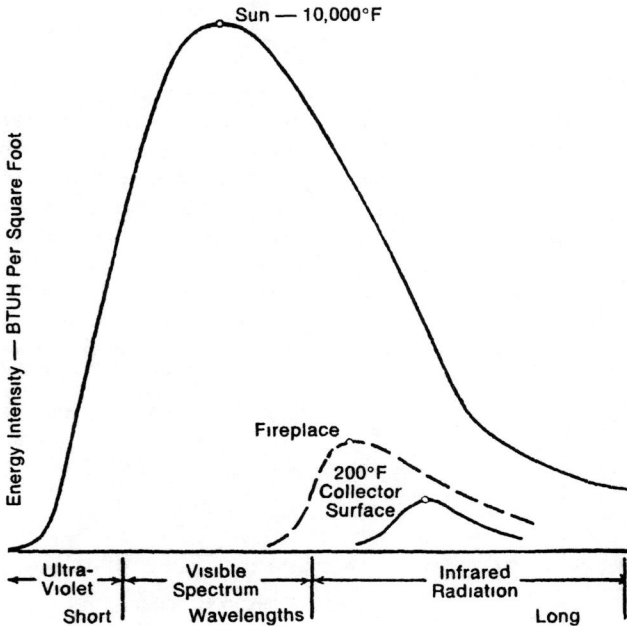

Fig. 3-5. Shift in wavelength toward infrared as temperature of surface decreases. Collector emits in infrared range which is absorbed and re-radiated by glass.

tempered in order to withstand abuse from wind, rain, hail, ice, air pollutants, and vandalism. A wire mesh can be used to protect the surface against flying objects but the efficiency of the collector will probably be lowered.

At the present time, glass is the principal material used to glaze flat plate collectors. However, a few varieties of specialized plastic materials can be used. The *second* cover may be made of heat resistant plastic.

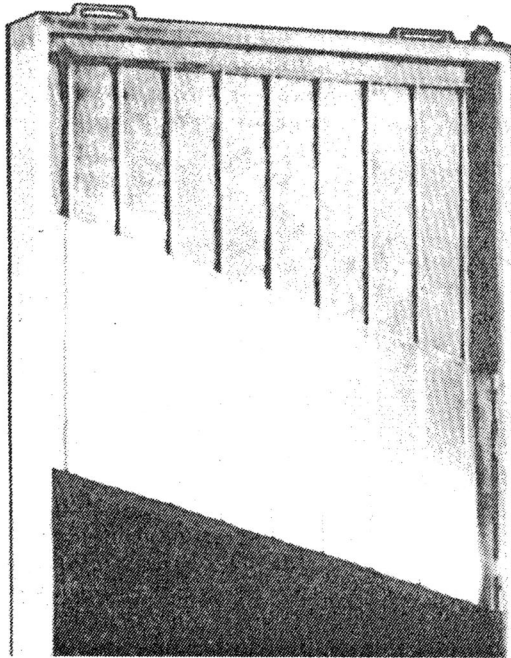

(Libby-Owens-Ford)

Another collector design uses copper tubes and a blackened aluminum sheet bonded to the tubes.

(Olin Brass)

One method of making an absorber panel is to bond two sheets of copper or aluminum together and then expand certain areas to form channels for fluid paths.

Space Between Glass Plates

The air space between glass plates must be effective in preventing excessive heat loss (from collector out to surroundings). Therefore, each enclosure between the plates and the absorber must have nearly an air-tight separation. This may be accomplished by installing flexible rubber mountings or using other sealing techniques.

An evacuated space between glass plates might prove very effective, but presently it is not practical in conventional sheet configuration. (Note: "Thermopane" or double glazed windows contain dry air or low conductance gas between panes—not a vacuum.)

Absorber Plate

Absorber plates are referred to as "blackbodies" because, in theory, they totally absorb all visible light radiation and the color black represents the lack of visible light reflection. There is no such thing as a perfect absorber or a surface which absorbs all the radiant energy falling upon it with none being reflected. Because about 40% of solar radiant energy is in the visible range (5% in ultraviolet, 55% infrared), the color of a surface can be important in determining the amount of energy absorbed. Figure 3-6 shows a comparison of surfaces colors and the percentage of solar radiation absorbed by each of them.

Figure 3-6

Note in Figure 3-6 that a black surface absorbs 92% of the radiant energy while a white surface only absorbs 40%. This is why a blacktop parking lot is so hot under foot and outdoor storage tanks are often painted white or are covered with a bright shiny surface to minimize heating the contents inside the tank. The heat absorption prin-

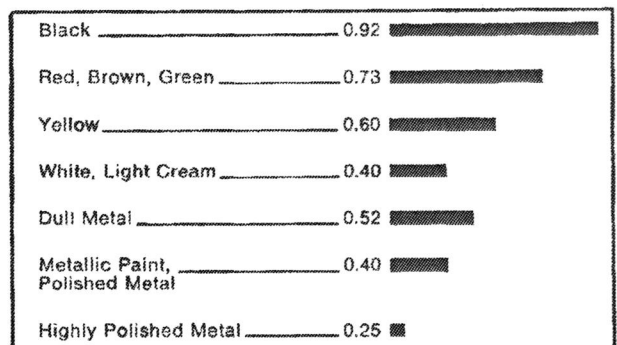

Black	0.92
Red, Brown, Green	0.73
Yellow	0.60
White, Light Cream	0.40
Dull Metal	0.52
Metallic Paint, Polished Metal	0.40
Highly Polished Metal	0.25

Fig. 3-6. Absorption of solar radiation by different surfaces.

ciple explains why absorber surfaces in a solar collector usually have a black finish.

The most universally used finish on absorber plates is flat black paint over a primer. This practice has been found to be very suitable for large temperature ranges and is very durable for long periods of time.

Factors other than color contribute to the operating efficiency of the absorber plate. While absorbing solar energy, the temperature of the absorber plate rises and the absorber itself becomes an emitter (sender) of radiation. To minimize the outward loss of heat by radiation from the hot absorber, glass plates are depended upon to absorb the longwave radiation. Remember that glass is highly resistant (opaque) to longwave or infrared radiation and that most plastics are not as effective as glass at absorbing radiation from the absorber. To increase the greenhouse effect, some absorbers are coated with materials other than paint which are called a *selective surface*. These special coatings impede the reradiation of infrared energy from the hot absorber and therefore reduce heat loss of the collector. As in most instances in applied technology, selective coatings have their drawbacks too. For one thing, they also affect to some extent the total energy absorbed by the surface; they are expensive, and their life expectancy is still undetermined. Therefore, black coatings are generally applied to the commonly used aluminum, copper, and/or steel plates when they are manufactured.

Other characteristics of absorber plates include: (1) absorptivity approaching 95%, (2) minimum thermal resistance between the plate and the transfer medium, (3) maximum thermal conduction between the plate and the transfer medium, and (4) emissivity of infrared rays from the plate to the transfer system.

Coating materials are manufactured for various purposes. The most universally used is flat black paint over a primer. This has been found to be very suitable for large temperature ranges and durable over long periods of time. Alternatives for paint, still being researched, are called selective surfaces. They contain many of the essential physical, optical, and thermal characteristics but have been considered (1) nondurable over a period of time, (2) high or speculative in cost, (3) generally unavailable because of the application process, (4) unproven in their performance, and (5) not that much better than paint when they are evaluated.

Heat Transfer Systems

Transfer systems refer to the method of moving a heat "pick-up" fluid through air or liquid collectors. These carry away the heat for delivery to other components for direct heating or for storage. Since the operation of the entire system will be discussed later, only the design of the heat transfer system involving the collector will be described here.

Air Collector System. In an air collector, air is forced through the collector by means of a conventional furnace-type blower. Figure 3-7 illustrates several different types of air collectors. The major differences exist in the design of the air heating space by having metal fins, black gauze or glass plates which increase the internal collector surface area with which the air makes contact and thus improves efficiency.

Diagram A illustrates absorber-to-air channel heat transfer with unrestricted air flow. The others, with the fins, gauze, glass plates or irregular metal surface, increase the metal-to-air heat transfer surface as well as create air turbulance to aid in conducting heat. Collectors of the styles B, C, D and E also impose greater resistance to air flow and thus require a more powerful fan capable of moving a greater number of cubic feet per minute (CFM) of air.

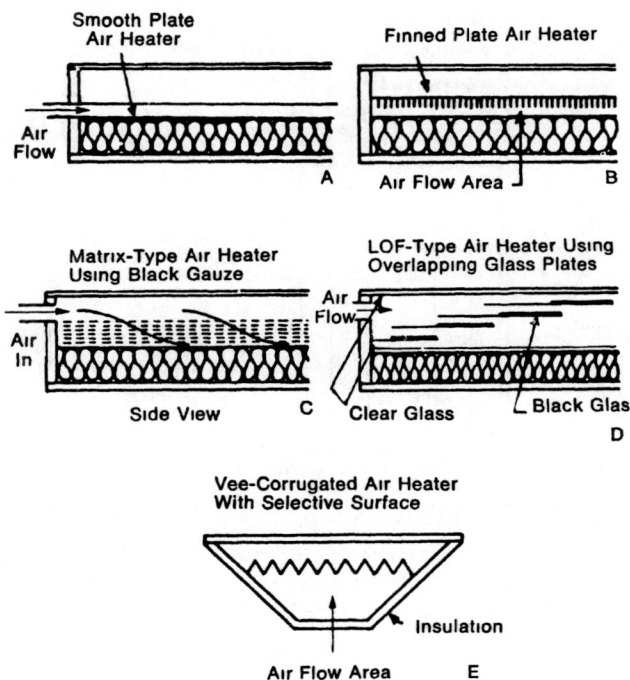

Fig. 3-7. Typical air collectors (cross section).

Serpentine
(A)

Grid
Direct Return
(B)

Grid
Reverse Return
(C)

Fig. 3-8. Basic flow patterns for liquid collectors.

In an air system, ductwork requires considerable space and must be well insulated. There are no freezing, corrosion, or structure damaging leak problems with air collectors.

Liquid Collector System. Moving a liquid through a collector presents a number of design problems. The collector design must provide for uniform flow through the collector by the specific way the tube configurations are attached to, or have been manufactured as, an integral part of the absorber.

The *serpentine* design (A) in Figure 3-8 is easier to construct but imposes high resistant to fluid flow. The *grid direct return* (B) design has balancing problems such as conventional hydronic system piping have. Valves and/or tube constrictions are used to equalize flow. The *grid reverse return* (C) design is probably the most efficient. In this design, the flow is evenly distributed through the tubes when the pressure drops through the headers is less than 10 percent of the drop in the tubes.

There are a number of tube designs currently being used in liquid transferring collector plates. Some styles of collectors are used for pressurized systems (rated at 125 psi maximum working pres-

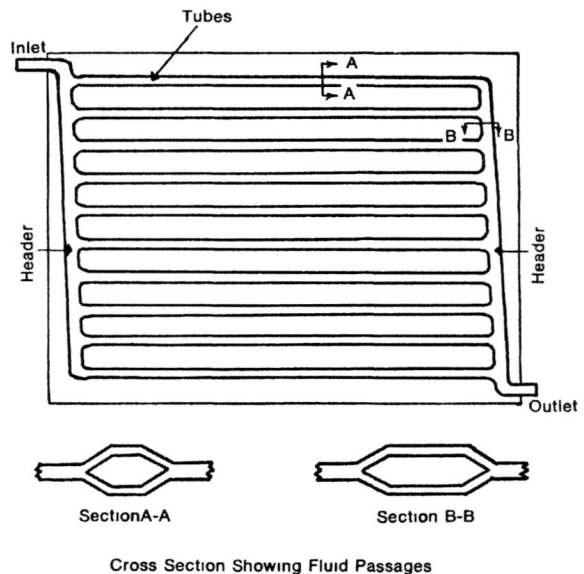

Cross Section Showing Fluid Passages

Fig. 3-9. In reverse return pattern, the pressure drop through the header must be less than 10 percent of the pressure drop through the tubes to provide even distribution of liquid flow.

sure) while others are for low pressure and thermosyphon systems (rated at 30 psi maximum working pressure). With the exception of Diagram I and J in Figure 3-10, the fluid travels *upward* as it heats. The Thomason System is a trickle type with water being heated as it flows, exposed to the sun, *down* through the troughs.

A critical problem in the manufacture of absorber plates is that of being sure the fluid tubes are thoroughly (and permanently) bonded to the plate, whether their passage way is above, within, or below the absorber plate. Otherwise, heat transfer from plate-to-tubes-to-fluid becomes impeded and results in lower collector efficiency. Some designs feature liquid paths formed as an integral part of the absorber.

The most critical concern with liquid collectors is with the possibility of freezing. This is particularly true when ordinary water is used as the circulating fluid. If the pipes leading to and from the collector freeze, and if the sun heats the absorber plate, the liquid trapped in the collector will get hot and possibly boil. As a result the thin metal tubing in the collector may rupture. In cold climates, an antifreeze solution or a special heat transfer fluid is used to prevent freezing in systems that do not feature a *drain-down* arrangement. There will be more discussion on this later.

Another problem with the liquid system is that the flow of liquids in the collector accelerates corrosion. Electrolysis can occur when dissimilar metals such as copper, aluminum, and steel are connected together. There are some special problems in using aluminum for pipes, plates, and other components. Ethylene glycol with special inhibitors must be used. Also, the reaction of copper and iron with aluminum causes corrosion of the aluminum. To counteract this cor-

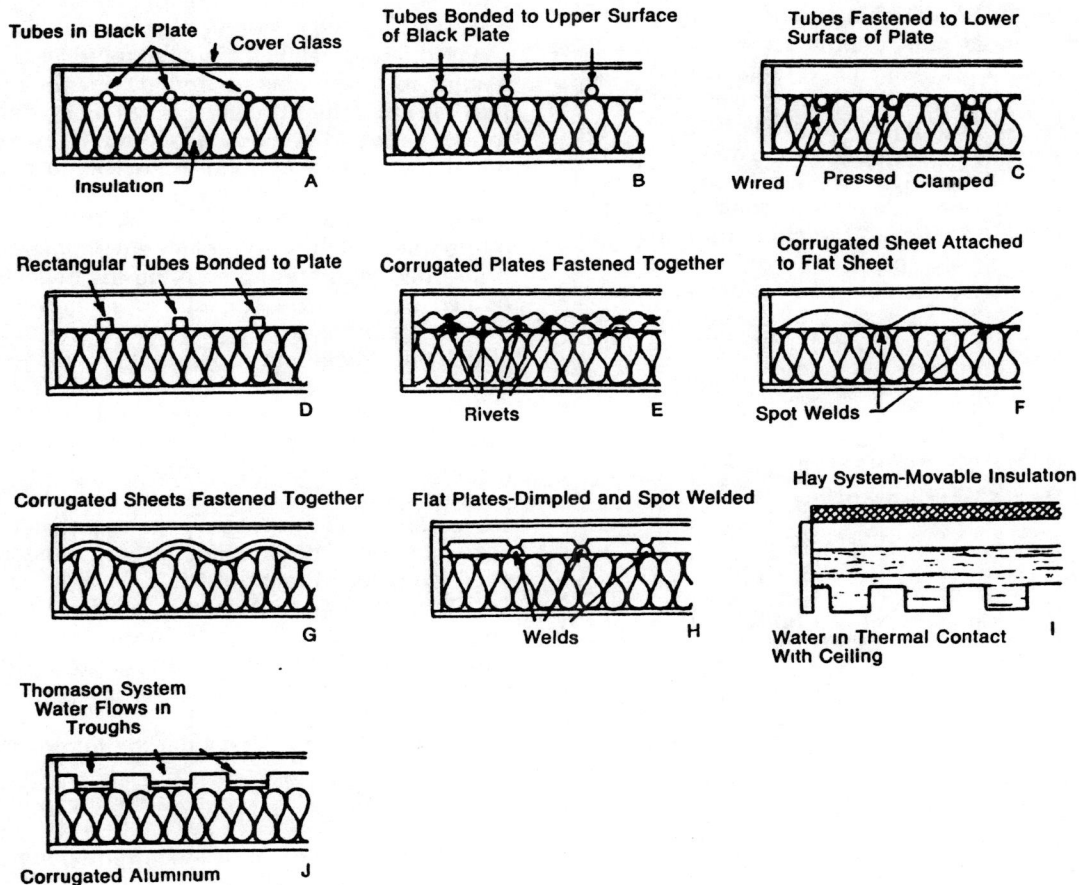

Fig. 3-10. Typical liquid collectors (cross section).

rosion, special precautions must be taken which are discussed in the Installation section.

Insulation

The purpose for insulation on the sides and back of the transfer system is to prevent heat from escaping through the frame. The insulating material should have low thermal conductivity, low bulk density, and high melting point. Insulating material should also be resistant to rot, weather, insects, and rodents. Various types of fiberglass and polyurethane foam meet these criteria.

Frame

The frame or box in which the transparent cover(s), absorber plate, transfer system and insulation are packed must be sealed against weather and climate. The frame must expand and contract with the rise and fall of external and internal temperatures. Such movement is necessary in order to maintain the necessary support for the internal components as well as the airtight seals between the various parts. Thus, frames may be made from galvanized steel, aluminum, or fiberglass. They may be built-in (attached directly to the roof framing material on the building) and actually become an integral part of the roof although this may pose repair and replacement problems. Free standing frames, attached to and elevated above roof, are also used, especially where the collector frames are installed on an existing structure. Frames are oriented with the inlet on the eave end of the collector and the outlet up toward the peak of the roof as illustrated in Figure 3-11.

HOW EFFICIENT IS A FLAT PLATE SOLAR COLLECTOR?

Evaluating the efficiency of a collector is a matter of determining how well the collector does what it is designed to do. From the discussions in this lesson, it can be determined that the efficiency of a collector is established by how much solar energy is transferred to the circulating fluid. That is:

$$\text{Efficiency} = \frac{\text{Btu/h Output}}{\text{Btu/h Input}}$$

Since there are many variables involved, determining a single point rating of a collector is not attempted at the present time. ASHRAE Standard 93-77 details methods of testing solar collectors and establishing collector efficiency.

Figure 3-12 shows a typical representation of a solar collector's instantaneous efficiency. Note that the collector efficiency is not a constant factor but varies with conditions. First, if the fluid inlet temperature and solar intensity are constant but the outdoor temperature decreases, the efficiency of the collector decreases. This is a result of greater heat loss in a colder environment. Also, if the fluid temperature entering the collector becomes warmer, the collector efficiency will decrease. In other words, the hotter the collector gets, the lower its operating efficiency will become.

The value of the slope (rise over run) of the efficiency curve and the value of the intercept (where curve meets vertical scale) are two very important characteristics of a solar collector. These values are used in all basic sizing procedures to determine how much collector area is needed to supply a given number of Btu's.

The slope and intercept are often listed in a collector manufacturer's literature under performance characteristics and the following mathematical terms:

$$F_R(\tau\alpha)_n \qquad \text{equals the intercept}$$

$$F_R U_L \qquad \text{equals the slope}$$

The value of each of these terms can be obtained directly from the efficiency curve whenever fluid temperature "in" is used in the term (Fluid

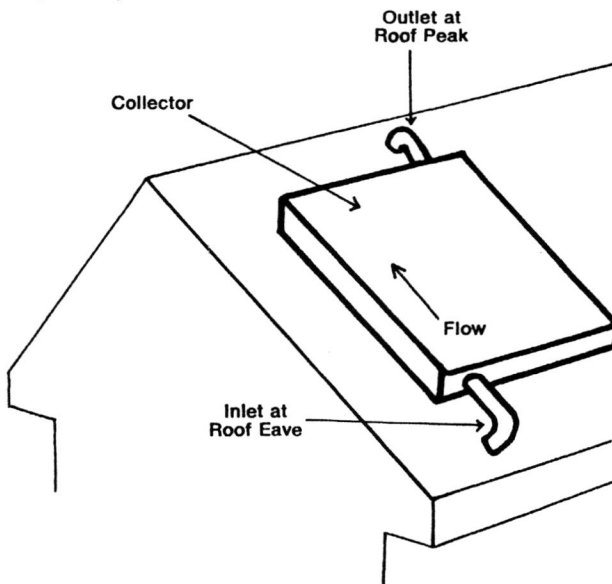

Fig. 3-11. Collector is mounted with flow direction up toward roof peak.

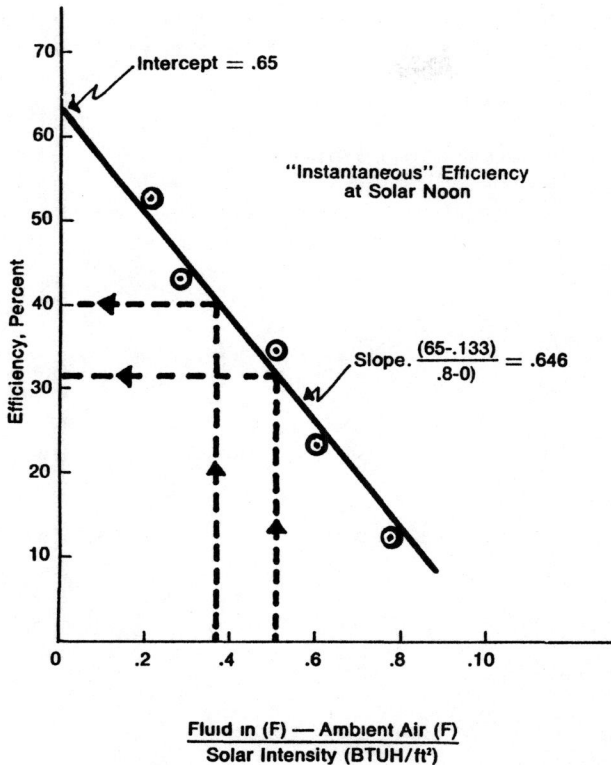

Fig. 3-12. Collector efficiency at solar noon.

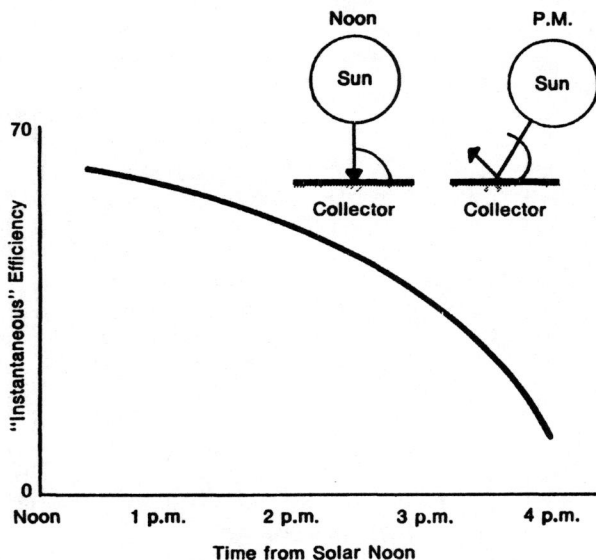

Fig. 3-13. Efficiency of fixed collector decreases as the sun "moves" across sky and the azimuth angle between sun and collector surface decreases.

in—Ambient air)/Solar Intensity as plotted along the horizontal scale in Figure 3-12.

For a fixed-flat plate collector, the efficiency becomes less as the sun moves across the sky and the angle between sun and collector surface decreases, as illustrated in Figure 3-13. This efficiency loss results from reduced interception and greater reflection.

Most data about collector efficiencies represent performance at or near solar noon when the sun is most nearly over the collector. Thus, an all-day or an *average* collector efficiency will be less than the published value.

In lieu of a standard rating, collector manufacturers should, at least, provide test results as conducted in conformance with ASHRAE Standard 93-77.

Transfer Fluid Flow Rates

There is a practical range of liquid flow rates through a solar collector. Increasing flow above a certain value does not measurably increase collector heat output, but circulation rates below a specified minimum will cause collector output to fall off significantly. This is roughly-analogous to water flow rates through hydronic baseboards, where a minimum flow for effective heat transfer is established, and a reasonable maximum value exists beyond which no real gains are accomplished. Typically, fluid flow is .02 gpm (gallons per minute) per square foot of collector.

For air collectors, a typical range is from 2-3 cubic feet per minute (cfm) per square foot. Both air and liquid collectors operate at similar temperature levels, with absorber plate temperatures typically at 150° F depending on the solar intensity.

SUMMARY

Solar collectors are the "heart" of the solar assisted heating system. The collection of heat depends on their efficiency and the amount of direct and diffuse solar radiation that strikes them. Flat plate collectors are designed and manufactured for both liquid and air heating systems, while concentrating collectors are made exclusively for liquids. Once the fluid is heated, it is transferred to the heat storage unit—the major topic to be studied in Lesson Four.

HEAT STORAGE

4

The sun is an *unlimited* and *variable* source of energy. In North America, only about 14 hours of sunlight can be expected in summer and perhaps 8 hours in winter, under ideal conditions. Local weather conditions (sunshine or clouds) will change the amount of solar energy actually received by the collector and transferred through the system to the interior of the structure.

With a heat supply that is both variable and interruptable, it is logical to collect solar energy when it is abundant, use it as needed, and store any extra heat for later use, either at night or on sunless days. In the early days of solar heating, the practice was to install a very large storage capacity that was planned to supply heat for several days without sun. Today, the trend is toard a more modest storage capacity.

CONCEPTS OF HEAT STORING MATERIALS

All substances—whether they are in a solid, liquid, or gaseous phase (form)—are capable of absorbing heat. Any given substance has a spe-

Fig. 4-1. Fiberglass tank for "low" temperature water storage. Residential needs would be satisfied by a much smaller tank.

cified relationship to water in its ability to absorb heat. Water has a *specific heat* of 1.0 Btu per pound for each degree of Fahrenheit rise in temperature. This means that a cubic foot of water weighing 62 pounds (or 8.34 lb/gallon) will have absorbed 62 Btu's if its temperature rises 1°F. For example, a 2000 gallon water storage tank with water heated from 90° to 100°F would absorb 2000 gal × 8.34 lb/gal × 1 Btu/lb × (100°F - 90°F) or 166,800 Btu's.

Rock, such as granite, is another common material that can be used for heat storage. The specific heat of rock is about 0.20 Btu/lb per °F. A cubic foot of lightly packed two-inch rocks weighs about 100 pounds. Therefore, a cubic foot of rocks would store 20 Btu's when raised 1°F.

These heat storage capacity specifications, for water or rock, help to explain the volumetric space requirements for fluid storage mediums. In a one cubic foot container, five times as many BTU's can be stored if water, rather than rock, is the storage medium. The solar heating system that functions by warming water or rock is classified as a *sensible heat* storage system.

The alternative to the sensible heat storage system is the *phase change* storage system. An example of this type of heat storing practice may be seen with paraffin wax. During the day, the wax melts as it absorbs solar heat. During the night, it cools and freezes as it releases heat to the air. Melting wax has about four times the heat absorbing qualities as water and, theoretically, could store a comparable amount of heat in ¼ the storage space required for water.

When the storage medium undergoes a change of phase: that is, changes from a solid to a liquid and back again, the term "heat of fusion" storage is sometimes applied. Almost any substance can be changed from solid-to-liquid-to-gas by the addition of heat as shown in Figure 4-2. In this diagram, the solid substance is ice. As heat is applied and absorbed by the ice, temperature is raised to 32°F whereby the ice melts. It absorbs 144 Btu's for each pound of melting ice. Further heating increases the temperature of the liquid from 32° to 212°F. At 212°F, a second phase change occurs as the water turns to steam—absorbing 970 Btu's per pound in the process. If the process were reversed, the same number of Btu's would have to be removed as the steam changes to water and then to ice.

With certain waxes and salts, however, the solid-to-liquid heat of fusion can be made to occur

4-1

at more elevated temperatures, between 90° and 120°F. Therefore, a larger amount of heat associated with phase change is gained or lost at a temperature suitable for direct use in space heating.

Phase change materials have some disadvantages. Substances such as waxes and salts are quite expensive at the present time. Also, many must be replaced regularly because of chemical decomposition. In addition, the design of phase change containers is complicated by the need for a large heat transfer surface. Hence, much of the space saving feature of using a phase change material is lost because the container must be enlarged for practical considerations.

The question of what storage medium to utilize is based on (1) the ability of a substance to accept heat, (2) the cost of the amount required for a given heat demand, and (3) the availability of the material. With this information in mind, the remainder of the lessons in this course will be limited to sensible heat storage practices using water and rocks.

CONCEPTS OF HEAT STORAGE SIZING

In the first section of this lesson, the nature of different heat storage materials was discussed. The purpose of that information was to provide an awareness of some of the possible mediums to use for storing heat. In this section the essential

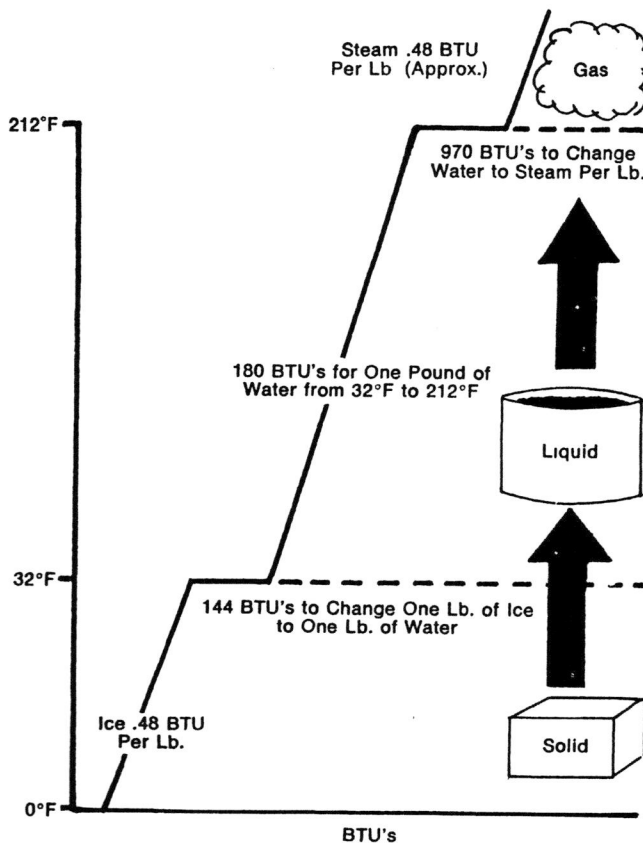

Fig. 4-2. Phase change from solid to liquid to gas.

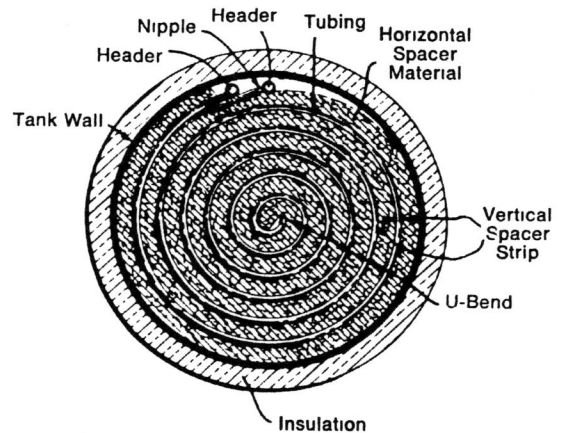

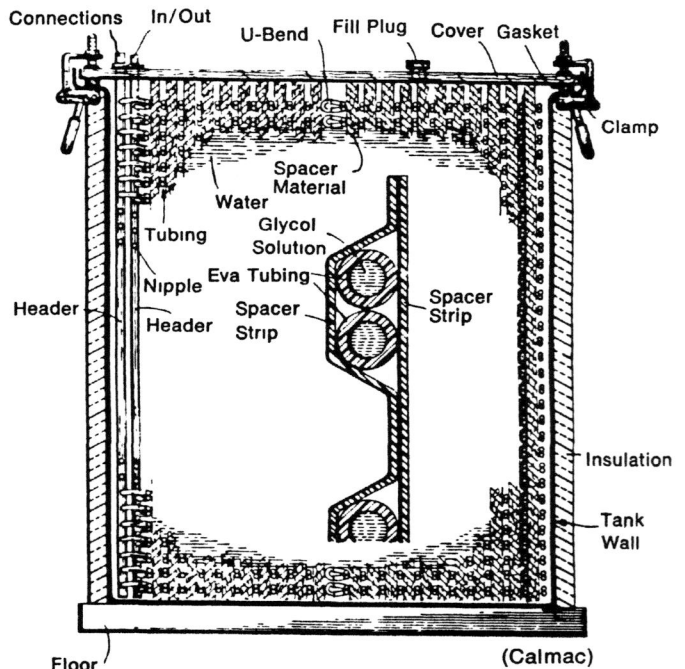

Fig. 4-3. Phase change storage unit. Plastic mat heat exchanger sits in container filled with an inorganic salt hydrate.

4-2

considerations of sizing air and liquid storage units will be explained as a preview for further discussion in Lesson Six.

Solutions to the problem of adequate heat storage relate to both the building's heat loss or "energy demand" and the amount of energy that the collectors can provide. In other words, the volume of space for storage depends on the size of the building and the size of the collectors.

Collector sizing will be treated in another lesson. At this point, merely appreciate the fact that, in the normal design process, the collector area required would be finalized first and storage selected would be based on collector output.

The interdependence of storage size upon collector size is illustrated in Figure 4-4. Time of day is represented on the horizontal scale and energy in Btu's per hour is provided in the vertical scale. The top horizontal line represents the heat loss of a building as might be calculated using *NESCA's Manual J* or *SMACNA's Load Calculation Guide*. The wavy horizontal line indicates the actual hourly heat loss for the hypothetical building on a particular day when the outdoor conditions were much milder than the design day. From just before 9 a.m. and just after 3 p.m., there are two curves for the amount of solar energy collected by two different sizes of collector arrays. The shaded area indicates the amount of energy

used to heat the house between 9 a.m. and 3 p.m. The net amount of collected energy available for storage is the clear area above the shaded region. The smaller collector obviously has less excess energy available for storage than the larger collector. This factor is always true, regardless of the size of the storage unit that is installed. In other words, a storage unit can be too big.

Figure 4-4 illustrates only one operating mode. There are other possibilities. On a very mild day, all the energy collected could go to storage and, conversely, on a near design day, there may be no excess collectable energy for storage.

A storage capacity that is too large could also affect the temperature of the storage medium, such as the water in a tank. There is a

Fig. 4-5. Small insulated water storage tank. Note thermometer in immersion well, pumps left and right sides and partially insulated piping.

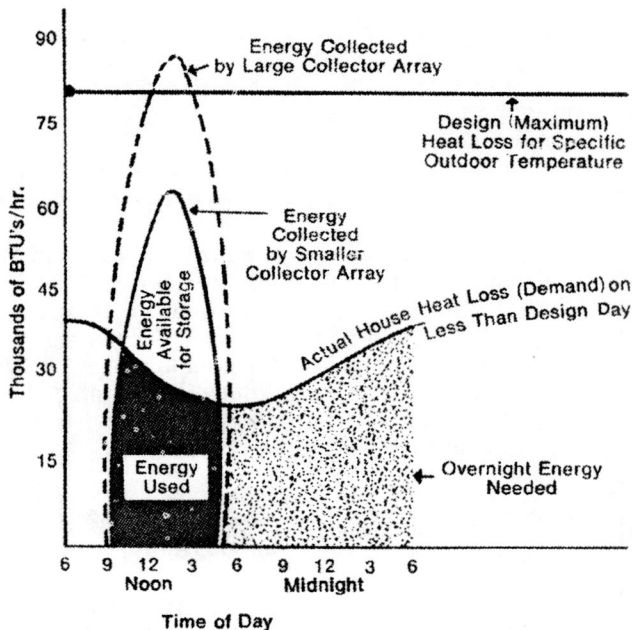

Fig. 4-4. Solar energy availability and needs.

minimum supply water temperature for effective heating with a fan coil, panel, or baseboard unit. An excessively large storage tank may take days to be heated to a useful temperature if the collector to storage capacities are extreme (small collector, large tank).

Conversely, a storage unit that is too small may become overheated and the operating temperature in the collector may be excessive. This would decrease the collector's efficiency.

Figure 4-6 represents a trend curve noting the effect of storage size on the contribution of solar heat to the total heat supplied to a building. With no storage, a solar system would, perhaps, contribute no more than 40% of the heating needs (assuming a fixed size of collector). The impact of providing a very large storage capacity would not significantly increase the percent contribution much above 80%, which could be achieved by a moderately sized storage unit. Cost consideration is still another factor that can limit the size of storage provided. Doubling the cost of the storage facility to gain two or three percent more solar contribution is *not* generally considered to be worthwhile.

What is a reasonable, moderate storage capacity for residential type solar heating installations? Designers do not all agree! But generally, storage of from one to two gallons of water per square foot of installed collector has been successful. For rock storage, from one-half to one cubic foot of rocks per sq. ft. of collector is quite effective. Proprietary solar assisted heat pump applications may require slightly different storage ratios, and the manufacturer must be consulted regarding these.

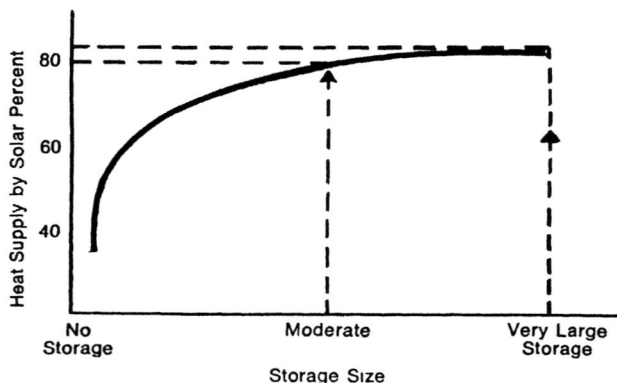

Fig. 4-6. Effect of storage size on heat contributed by solar energy.

The other side of the collector storage problem is—as was mentioned—the thermal size of the building.

To size conventional heating equipment, a design heat loss calculation is made. This is based on local design outdoor conditions and recommended indoor design temperature. The outdoor design temperature in a given locality is not the lowest ever recorded temperature but, rather, a somewhat higher temperature based on frequency of occurance. For example, Figure 4-7 illustrates a section of the ASHRAE Table of climatic data for the United States and Canada. (A full listing is included in the SMACNA Installation Standards). Note that three outdoor temperatures are listed for each city—median of extremes, 99% and 97½%. Median value is the middle value of the coldest temperature recorded each year for 30 years. The 99 and 97½ percent temperatures indicate the lowest temperature likely to occur for either 99 or 97½ percent of the total winter operating hours for December, January and February. In other words, out of 2160 hours in those three months, the temperature might fall below the listed value for each city for 21.6 hours for 99 percent and 543 hours for 97½ percent columns.

Since a design heat loss calculation is an estimate of the maximum or near maximum load of a structure in terms of Btu's per hour, it cannot be used to estimate the overnight, daily or monthly Btu requirements of the building. For example, a house with a design heat loss of 50,000 Btu will not require 50,000 × 24 hours or 1,200,000 Btu's per day unless the outdoor temperature for that particular day was at the design value for 24 hours. Since outdoor temperature is usually a variable through the day, a mean (average) daily temperature must be used to estimate daily Btu requirements.

Mean daily temperature is ½ the sum of the maximum and minimum temperature for a specific day, see Figure 4-8. For example, assume that in a 24 hour period, the outdoor temperature ranged from a high of 40°F to a low of 18°F. The mean daily temperature would be 40 + 18 divided by 2, or 24°F. The estimated daily Btu requirements are as follows:

$$\frac{\text{design heat loss}}{\text{design temperature difference}} \times \frac{\text{Mean daily temperature difference}}{} \times 24\text{ hrs.}$$

= daily Btu requirements

If the indoor temperature is 70°F, and the outdoor design temperature is 0°F, then using a 50,000 Btu house, the formula is:

$$\frac{50,000}{(70 - 0)} \times (70 - 24) \times 24 = 788,571 \text{ Btu's per day}$$

As is generally acknowledged, almost any building has internal heat released from lights, cooking, and other activities, there is also some solar heat gain directly through windows. In an effort to account for this heat contribution, and to more precisely estimate what actual Btu's the

Climatic Conditions

Col. 1 State and Station[h]	Col 2 Latitude[i] ° '	Col. 3 Elev,[d] Ft	Winter			Col. 5 Coincident Velocity Wind
			Col 4			
			Median of Annual Extremes	99%	97½%	
NORTH DAKOTA						
Bismarck AP	46 5	1647	—31	—24	—19	VL
Devil's Lake	48 1	1471	—30	—23	—19	M
Dickinson AP	46 5	2595	—31	—23	—19	L
Fargo AP	46 5	900	—28	—22	—17	L
Grank Forks AP	48 0	832	—30	—26	—23	L
Jamestown AP	47 0	1492	—29	—22	—18	L
Minot AP	48 2	1713	—31	—24	—20	M
Williston	48 1	1877	—28	—21	—17	M
OHIO						
Akron/Canton AP	41 0	1210	— 5	1	6	M
Ashtabula	42 0	690	— 3	3	7	M
Athens	39 2	700	— 3	3	7	M
Bowling Green	41 3	675	— 7	— 1	3	M
Cambridge	40 0	800	— 6	0	4	M
Chillicothe	39 2	638	— 1	5	9	M
Cincinnati CO	39 1	761	2	8	12	L
Cleveland AP	41 2	777r	— 2	2	7	M
Columbus AP	40 0	812	— 1	2	7	M
Dayton AP	39 5	997	— 2	0	6	M
Defiance	41 2	700	— 7	— 1	1	M
Findlay AP	41 0	797	— 6	0	4	M

Fig. 4-7. Climatic conditions for North Dakota and Ohio from SMACNA standards.

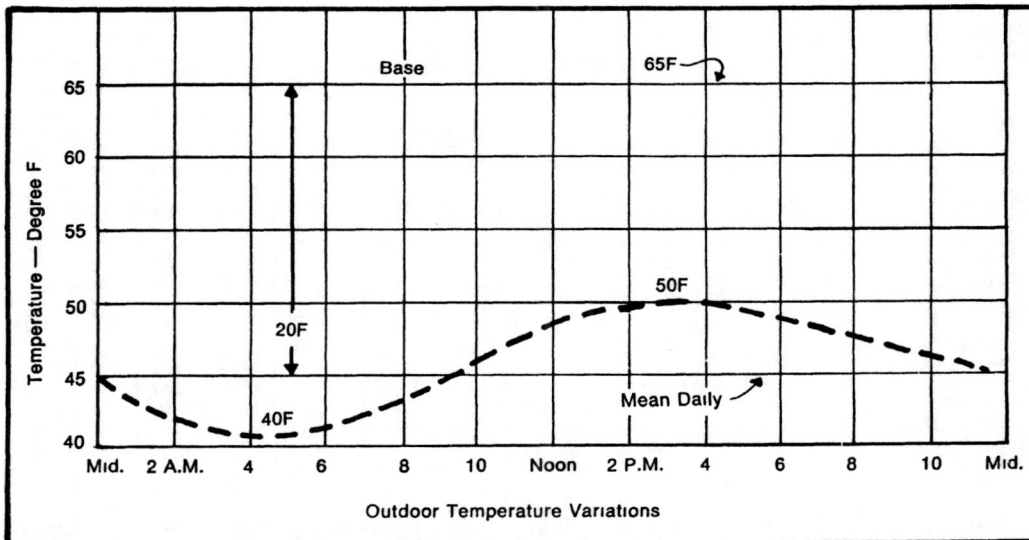

Fig. 4-8. Idealized outdoor temperature fluctuations throughout the day. Mean daily temperature is half the sum of the maximum and minimum temperature for a specific 24 hour period.

heating system must supply, the mean daily temperature difference is usually adjusted by changing the indoor temperature from 70°F to 65°F. Thus, the formula would be:

$$\frac{50,000}{(70-0)} \times (65-24) \times 24 = 702,857 \text{ Btu's per day}$$

The value (65-24) is commonly called Degree Days. Monthly and yearly Degree Days have been published for many years for given cities. A sample tabulation is shown in Figure 4-9. (Full listing is provided in the SMACNA Installation Standards.)

If, for example, the hypothetical building was located in Dayton, Ohio, which has 5622 annual degree days, the estimated annual energy requirements would be:

$$\frac{50,000}{(70-0)} \times 5622 \times 24 = 96,377,140 \text{ Btu's per season}$$

If an estimate is made of the January energy requirement for the building, use the January degree days for Dayton which are 1097. The estimated January load would be:

$$\frac{50,000}{(70-0)} \times 1097 \times 24 = 18,805,714 \text{ Btu's for January}$$

As a first approximation, it is also possible to divide the January total by 31 days to arrive at a "typical" day's energy need in January. Thus:

$$\frac{18,805,714}{31} = 606,636 \text{ Btu's per January day}$$

If the solar collectors operate for 8 hours, then 16/24 of the all day load would be approximately the Btu requirement needed from storage —provided a full 16 hour storage requirement was expected. Thus:

$$\frac{16}{24} \times 606,636 = 404,424 \text{ Btu's overnight load}$$

These numbers are approximations. In fact, the degree day concept itself for annual analysis has come under criticism because it was developed in the 1930's. Modern construction has apparently altered the 65°F base. Some designers apply corrections to the degree days to compensate for modern construction. Table 1 lists currently used corrections. For our example, Dayton has a 0°F design temperature, the correction would be 0.71. Hence,

$$404,424 \times 0.71 = 287,140 \text{ Btu's overnight load}$$

and the amount of energy the collectors must supply in excess of the previous day's needs. How much storage would be required to satisfy the estimated overnight load?

If water was used and a 30°F rise in storage temperature occurs, then:

Average Monthly and Yearly Degree Days

State	Station		Avg. Winter Temp.ᵈ	July	Aug.	Sept.	Oct.	Nov.	Dec.	Jan.	Feb.	Mar.	Apr.	May	June	Yearly Total
	New York (Kennedy)	A	41.4	0	0	36	248	564	933	1029	935	815	480	167	12	5219
	Rochester	A	35.4	9	31	126	415	747	1125	1234	1123	1014	597	279	48	6748
	Schenectady	C	35.4	0	22	123	422	756	1159	1283	1131	970	543	211	30	6650
	Syracuse	A	35.2	6	28	132	415	744	1153	1271	1140	1004	570	248	45	6756
N.C.	Asheville	C	46.7	0	0	48	245	555	775	784	683	592	273	87	0	4042
	Cape Hatteras		53.3	0	0	0	78	273	521	580	518	440	177	25	0	2612
	Charlotte	A	50.4	0	0	6	124	438	691	691	582	481	156	22	0	3191
	Greensboro	A	47.5	0	0	33	192	513	778	784	672	552	234	47	0	3805
	Raleigh	A	49.4	0	0	21	164	450	716	725	616	487	180	34	0	3393
	Wilmington	A	54.6	0	0	0	74	291	521	546	462	357	96	0	0	2347
	Winston-Salem	A	48.4	0	0	21	171	483	747	753	652	524	207	37	0	3595
N.D.	Bismarck	A	26.6	34	28	222	577	1083	1463	1708	1442	1203	645	329	117	8851
	Devils Lake	C	22.4	40	53	273	642	1191	1634	1872	1579	1345	753	381	138	9901
	Fargo	A	24.8	28	37	219	574	1107	1569	1789	1520	1262	690	332	99	9226
	Williston	A	25.2	31	43	261	601	1122	1513	1758	1473	1262	681	357	141	9243
Ohio	Akron-Canton	A	38.1	0	9	96	381	726	1070	1138	1016	871	489	202	39	6037
	Cincinnati	C	45.1	0	0	39	208	558	862	915	790	642	294	96	6	4410
	Cleveland	A	37.2	9	25	105	384	738	1088	1159	1047	918	552	260	66	6351
	Columbus	A	39.7	0	6	84	347	714	1039	1088	949	809	426	171	27	5660
	Columbus	C	41.5	0	0	57	285	651	977	1032	902	760	396	136	15	5211
	Dayton	A	39.8	0	6	78	310	696	1045	1097	955	809	429	167	30	5622
	Mansfield	A	36.9	9	22	114	397	768	1110	1169	1042	924	543	245	60	6403
	Sandusky	C	39.1	0	6	66	313	684	1032	1107	991	868	495	198	36	5796
	Toledo	A	36.4	0	16	117	406	792	1138	1200	1056	924	543	242	60	6494
	Youngstown	A	36.8	6	19	120	412	771	1104	1169	1047	921	540	248	60	6417

Fig. 4-9 Degree days from SMACNA standards.

$$287,140 = 8.34 \text{ lb/gal} \times 1 \times 30°\text{F} \times \text{gallons}$$

$$\text{or gallons} = \frac{287,140}{8.34 \times 1 \times 30} = 1148 \text{ gallons}$$

If rock storage is required and a 70°F rise in storage temperature occurs, then:

$$287,140 = 20 \text{ Btu/cu ft F} \times 70°\text{F} \times \text{cubic foot}$$

$$\text{or cu ft} = \frac{287,140}{20 \times 70} = 205 \text{ cu. ft. of rocks}$$

As described in a later lesson, it may not be economical to provide complete overnight storage. At this point, however, one should have an understanding of the calculations required to determine building needs and storage capacity.

PLACEMENT OF HEAT STORAGE UNITS

Solar heat storage units may be placed either inside or outside the building they are serving. They can be located either below grade or above grade. The rationale for a storage unit location is that it be placed in a *low value* area of the structure or building site.

Units placed below grade inside the building could be in a basement or crawl space. This area would be close to most of the other components in the heating system. It would be protected from moisture and cold. Any excessive radiation energy coming in and/or heat loss radiating from the storage container could be directed into the heated space as needed, or it could be vented to the outside through a duct system.

A low value space should also be sought when placing the heat storage unit above grade inside the structure such as a small room or attic. Because of the weight of the container and the storage media (pebbles or water), extensive foundation reinforcement would be necessary. In one case, to serve the needs of a *thermosyphon system* used primarily to service household requirements for domestic hot water, the heat storage tank must be about two feet above the collector. That would mean using an inconvenient, but low value, attic installation.

Units placed outside and above ground would be exposed to whatever minimum temperature that would be reached in a given geographical location. They would also create a problem aesthetically.

Units placed outside and below ground pose some problems also. They would need to be buried below the frostline so that they would not be affected by pressure from frozen soil. Water or air leaks would be difficult to locate. Digging may

be required before servicing unless the container was in some sort of a pit. In this case, an access cover would provide a way in to the storage unit.

There are problems of storage unit placement associated with each type of solar heating system. Not only is this true for the architects and contractors when constructing new facilities but, in addition, inconvenient when existing buildings are retrofitted with solar heat. Some previously used living areas must be converted to non-usable space to accommodate heating system components.

ADVANTAGES AND DISADVANTAGES FOR STORING HEATED LIQUIDS

There are advantages and disadvantages to each and every type of solar heating system, regardless of the type of fluid (liquid or air) for which the system was designed. Advantages are that: (1) water is a relatively abundant and inexpensive medium, (2) water absorbs and emits heat readily, and (3) space requirements for water storage unit placement is the least for any of the common heat storage mediums used presently.

However, there are some disadvantages. First, the system must be insulated against freezing. (It may be too costly to use antifreeze for all the water in storage). Second, boiling may occur in the collector, and a ventilating valve may have to be installed to remove steam pressure if overheating occurs. Third, circulating the water throughout the system would require a pump and the system would be vulnerable to leakage. This would create the need for periodic inspections, troubleshooting, and maintenance. A fourth problem is that components made of aluminum, copper or brass, and iron which are threaded or joined directly to each other will cause an electrolytic action. Such action will form initially at the connection, spread through the components, and reduce the flow rate by constricting the pipe diameter as illustrated in Figure 4-10

Over a long period of time, corrosion can cause total erosion of pipe or fitting, and a leak will occur. As a result, pressure in the system would drop, antifreeze and inhibitors would be lost, and the system would stop functioning.

The rate of corrosion can be affected in two ways: (1) by the amount and type of minerals in the liquid, and (2) by the temperature of the fluid, since heat increases the rate of corrosion. One form of controlling corrosion is to add a corrosion inhibitor into the system. Another procedure is to use lengths of plastic pipe or dielectric unions with plastic bushings between the incompatible metal components.

A fifth problem for a liquid storage system is accessability. Buried or enclosed storage tanks and/or pipes would be difficult to inspect for leakage. If a pipe or tank froze and ruptured, or if chemical corrosion caused a failure in the system, repairs could be extensive. This aspect applies not only to the replacement of various components of the solar heating system, but also to the possible need to redecorate walls, ceiling, or floors where damage may occurred.

WATER TANK MATERIALS

There are several different materials commonly used for liquid storage systems. Concrete, steel, fiberglass, and butyl-lined concrete block are four of the materials frequently used. Selection should be made based on the required capacity and location of the storage unit. Figure 4-11 represents a typical tank installation.

Concrete Tanks

Concrete tanks are probably the most durable. These units cannot corrode or be punctured. A waterproof sealer on the inside surface or a waterproof liner will free the container from possible seepage problems. Another engineering problem to contend with is the weight and size of the tank to be installed. A crane or other heavy equipment is needed when the tank is set in place. This is particularly difficult when placing a tank in the basement of an existing structure. Watertight pipe connections are also difficult to achieve without using a high quality caulking compound such as silicone.

Fiberglass Tanks

A fiberglass tank especially designed for use with high temperature liquids is probably the most advisable system for water storage. The tank must be of a quality for protection against rupturing at a high temperature, such as the 212°F boiling point of water. Tanks used for gasoline storage may not meet this requirement. Care must be taken to determine safe pressure and temperature limits for the tanks. The tanks must be well insulated after they are installed, and must be checked for leaks if the solar heating system is to operate satisfactorily.

Glass-Lined or Galvanized Steel Tanks

Glass-lined or galvanized tanks have been found extremely serviceable for many years. Galvanized tanks, however, may become corroded and inserviceable before those which are glass-lined. The glass-lined tanks may have their expected years of usage shortened by careless installation. Excessive pressure in securing connections can crack or break the inflexible glass and allow more rapid deterioration. It may be advantageous to use several smaller tanks because they require less floor space in a basement and in a retrofit, smaller tanks may be installed more readily.

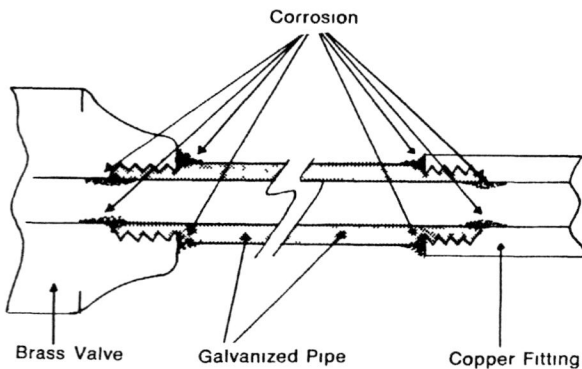

Fig. 4-10. Electrolytic corrosion.

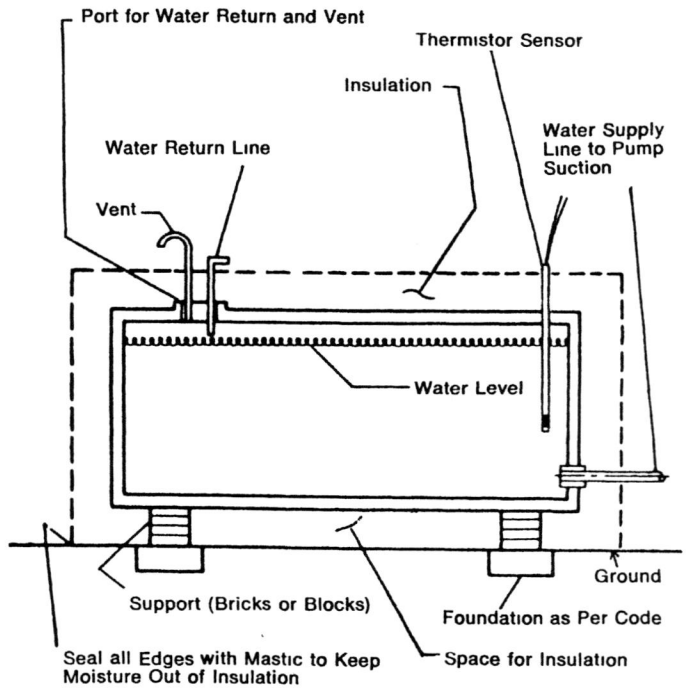

Fig. 4-11. Typical above ground liquid storage tank installation.

ADVANTAGES AND DISADVANTAGES OF PEBBLE-BED HEAT STORAGE UNITS

Heat storage units for air-circulating solar heating systems are called pebble-beds. They contain a large volume of pebbles (granite or other clean crushed rock about ¾ to 1½ inches in diameter). During the heat circulating cycle, air is forced through the inlet; then it filters through the pebbles and into ducts, where it is transferred to the space to be heated. When the rocks need to be recharged with heat, the air movement is reversed and hot air is drawn in through the outlet and heats the pebbles. A four foot deep bed of pebbles is considered the maximum for a minimum air pressure drop between the inlet and outlet.

Pebble-bed containers have been found very efficient. They are considered by many to be better than the water and/or phase change systems. They are simple to construct by either (1) laying cement blocks and filling the holes with concrete; (2) pouring concrete into a form; or (3) fabricating them from wall stud material, plywood, and insulation. The pebbles should be hand shoveled or chuted into the unit to insure even distribution and reduce packing or side wall stress. For example, a cylinder three feet by eighteen feet holds about 12,000 pounds of carefully sized crushed rock.

They have a high level of *thermal stratification*. Heat inside the container exists in layers. When the pebble bed is being charged with heated air from the collectors, most of the heat is transferred to the rocks within a few feet of the bed and the air leaving the bottom of the storage bin is essentially at 70°F. A representative temperature profile in a storage bin is shown in Figure 4-12. This profile obviously changes with the length of the charging time.

When flow is reversed to extract heat from storage, the hot end of the bin heats the leaving air to within a few degrees of the actual rock temperature for effective utilization of stored energy.

Cubic pebble-beds (Figure 4-13) are preferred because heat has a natural tendency to rise. This style of unit is normally about five feet high. Due to the weight of the container and its contents, these units must have rather extensive foundations so that the force of the weight will not cause the unit to crack. If a space three feet high is the maximum available (in a crawl space, for example), then a box using a horizontal air flow as shown in Figure 4-14 is adequate.

SUMMARY

Lessons Three and Four have presented the two main components of the primary circuit in a solar heating system. The collector, regardless of the type (liquid or air), functions by collecting the

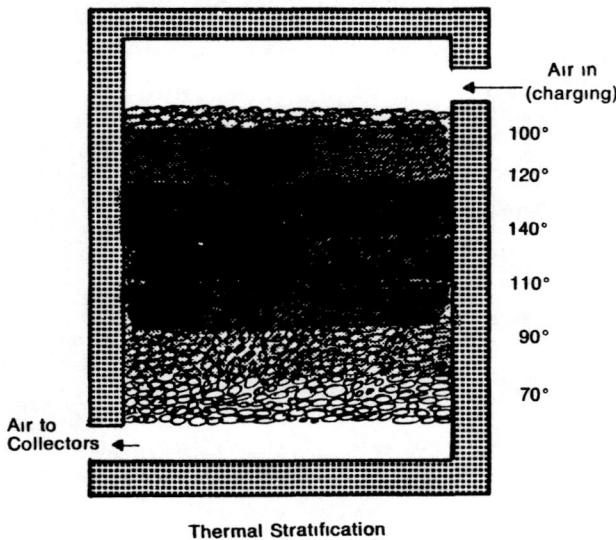

Air in (charging)

100°
120°
140°
110°
90°
70°

Air to Collectors ←

Thermal Stratification

Fig. 4-12. Temperature "layers" in pebble bed storage.

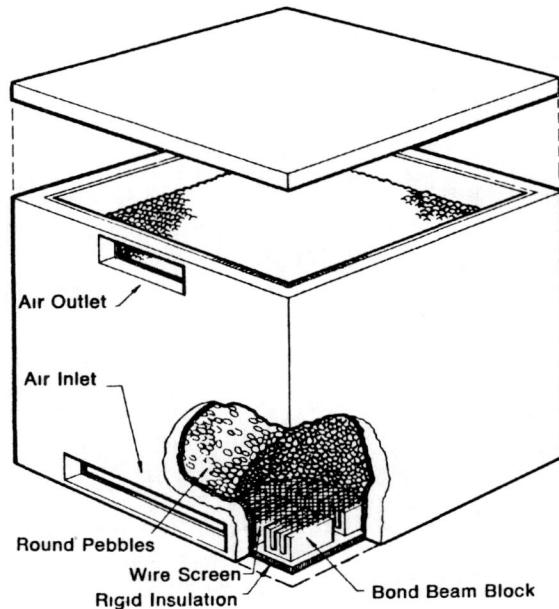

Air Outlet

Air Inlet

Round Pebbles

Wire Screen
Rigid Insulation

Bond Beam Block

Fig. 4-13. Cube type pebble-bed storage.

4-9

solar radiation from the sun. A liquid or air is circulated through the collector and absorbs heat energy which is transferred to the storage unit. The heat is stored until needed. When heat energy is needed, it is distributed throughout the space to be heated by a control system to be discussed in Lesson Five.

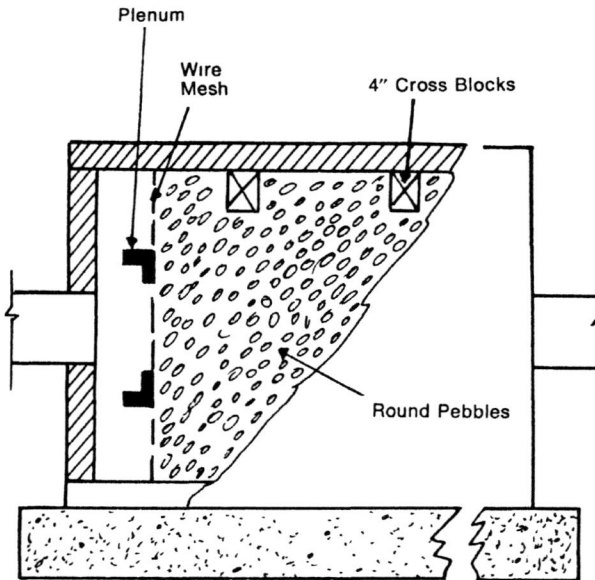

Fig. 4-14. Horizontal pebble-bed storage.

CONTROL DEVICES AND SPECIALTY ITEMS

In addition to the major items discussed in the previous lessons, there are a number of smaller but equally important components in a solar heating system. This lesson will be concerned with pumps, fans, and heat exchangers, and with the identification of the control devices and hydronic specialty items needed to operate a solar system safely and automatically.

Liquid System Pumps

Fluid movement is caused by electrically operated centrifugal pumps in a liquid-circulating closed or open loop system. Specifications for liquid-circulating pumps are based on the gallons per minute (gpm) to be circulated and the horsepower rating of the motor. The typical pump is of the type shown in Figure 5-1.

Pumps are placed in series within the collector loop. They must be capable of circulating the properly prescribed flow rate in gallons per minute through the collector to maximize the efficiency in collecting energy. When operating, the pump's *head pressure* (which is also called discharge pressure) must be sufficient to provide the proper flow to overcome resistance in the piping as illustrated in Figure 5-2. This resistance to flow is caused by the friction of the liquid as it passes through the collector, piping, pipe fittings and various components.

Figure 5-2 shows the relationship between pump pressure and pump capacity in gallons per minute for both a "large" and "small" pump. Note that pressure is given in terms of feet of water. One foot of water is equal to 0.43 psi (pounds per square inch). In manufacturer's literature, the pressure drop (resistance) for various components, collectors, coils, valves, piping, etc., will be given in either psi or feet of water. The important characteristic shown in the figure is that pump capacity decreases as the resistance or head against which the pump must operate increases. (Compare 22 gpm at 6 feet head vs only 12 gpm at 10 feet.) The resistance or pressure drop imposed by the piping, fittings, and other components must be determined before a pump can be chosen that will deliver the proper gpm at the estimated operating head pressure.

Collector loops vary in configuration, but in terms of pumping requirements, they are either a closed loop filled with a nonfreezing liquid (antifreeze) or simply a loop that completely drains down plain water each time the pump completes an operating cycle. If the loop remains filled with liquid at all times, then pump pressure requirements are minimal. Only sufficient pressure to overcome flow resistance is necessary, since the supply lift (elevation) to the collectors is "counter balanced" by a return drop. In the case of a drain down arrangement, the pump must lift the liquid to the top of the collectors at pump start-up and must overcome flow resistance once the loop is recharged (filled). Therefore, a circuit using draindown for freeze protection will usually require a larger pump and consume more energy while pumping than a closed antifreeze filled loop. Thus, it is generally less cost effective.

(ITT Bell & Gossett)

Fig. 5-1. Typical circulating pump.

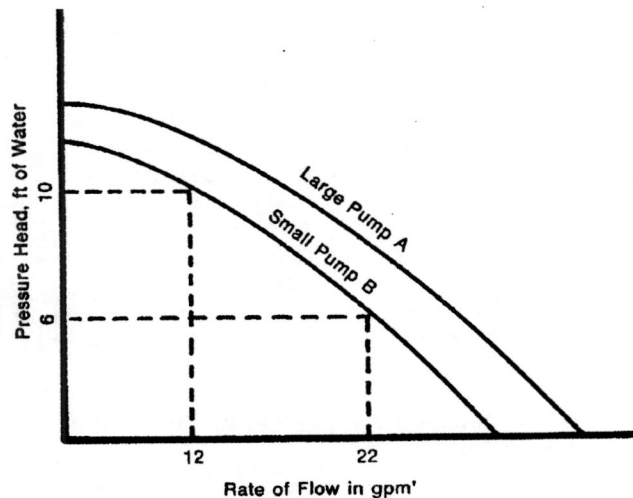

Fig. 5-2. Pump characteristics.

Pumps used in collector loops must be able to withstand both high pressure (to 150 psi) and high temperature (200°F. or even higher) and can be manufactured of a variety of materials. Generally, the construction material specified refers only to the parts that come in contact with the liquid being pumped. Thus, a bronze-fitted pump would have a bronze impeller, shaft sleeve, and wear rings. All-bronze, all-iron and all-stainless steel construction are also available. Manufacturers usually state uses for their various pumps, but in most cases, pumps used to pump water in systems open to the atmosphere (drain-down systems) must be bronze or stainless to minimize the corrosive effects of high oxygen content water. In closed loops, an iron pump is usually adequate.

Two speed pumps may be used in the collector loop. They can regulate the liquid temperature rise in the collector to optimize collector efficiency. When the solar radiation is high, the pump would transfer heat rapidly on the high speed setting. During times of low insolation, the slow speed would be more effective in circulating the heat energy.

Liquid System Heat Exchanger

One integral component of the liquid system collector loop is the heat exchanger. It will be identified in this section and explained in greater detail in a later lesson. A shell and tube heat exchanger is the most common type manufactured at this time.

The typical shell and tube heat exchanger is a heavily insulated tank containing a liquid heat transfer medium (water) and internal piping. A heated liquid containing antifreeze, transferred from the collector, is usually circulated through the shell. Water is circulated through the piping of the heat exchanger and is piped to the domestic hot water or auxiliary heating unit.

The purpose of this component is to reduce the quantity of antifreeze that would be needed if transfer fluids were exposed to outside temperatures while circulating through the flat plate collector mounted on the roof. By using a heat exchanger, only a small portion of the total number of gallons of water needed to transfer and store heat energy would have to be protected from freezing the fluid in the collector loop. A considerable amount of money can be saved in this manner. Figure 5-3 illustrates the function of the heat exchanger.

The shell of the heat exchanger contains the heat transfer medium (antifreeze) and circulates the collector-heated liquid around the piping inside the exchanger. The piping becomes hot. The heated pipe heats the water to provide hot water for space heating or storage.

Air System Blowers

In an all-air system, centrifugal blowers are used to circulate air through the collectors, heat storage unit, or directly to the conditioned spaces. Conventional warm air heating blowers can be used, since operating temperatures are essentially the same as found in conventional warm air systems. However, the system's operating pressure may be higher because of increased airflow resistance. (See Figure 5-4)

In any discussion of fans, it is always necessary to refer to three aspects of pressure—*static*, *velocity*, and *total*. So that there will be no doubt as to the meaning of these terms, these important parameters shall be defined.

Static Pressure (SP) is the pressure inside a container, for example, an inflated balloon or tire. It has no relationship to air motion, but it can exist

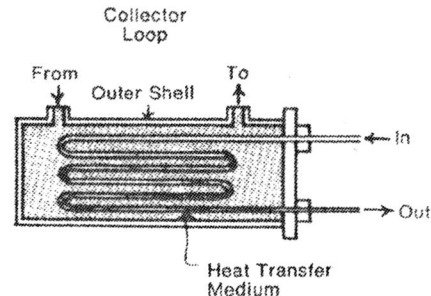

Fig. 5-3. Shell and tube heat exchanger. 1-1 schematic above, cutaway below.

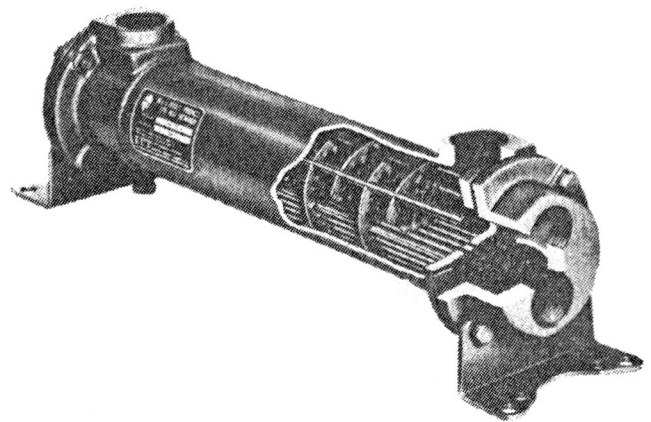

(ITT Bell & Gossett)

inside a duct where there is air motion. If the static pressure inside a closed container is below atmospheric pressure, the container will tend to collapse. If it above atmospheric, the container will tend to burst.

Velocity Pressure. (VP) is the pressure required to accelerate air from a state of rest to the velocity observed. Velocity pressure varies as the square of the air velocity, hence mathematically:

$$VP = KV^2$$

where

VP = velocity pressure
V = air velocity
K = a constant to account for dimensional units and gravity.

Normally, velocity pressure is measured in inches water gauge (WG), although feet WG, inches HG (mercury), or psi (pounds per square inch) all would be equally possible and valid.

Belt Driven Blower

Direct Drive Blower

Fig. 5-4. Two models of centrifugal blowers.

Total Pressure (TP) is the sum of the static and velocity pressures. Thus:

$$TP = SP + VP$$

where

TP = total pressure, in. WG
SP = static pressure, in. WG
VP = velocity pressure, in. WG

For low pressure duct systems, static pressure is usually specified for design purposes, since velocity pressure is a small value and can usually be ignored. All-air solar heating systems are usually designed using low pressure duct systems.

Figure 5-5 shows a fan performance curve plotting air delivery capacity in cubic feet per minute (cfm) versus static pressure in inches of water. As indicated earlier for pumps, the fan curve indicates a decrease in capacity as the operating pressure increases. A second curve, which is called the *duct system resistance curve*, is also shown. This curve is determined by estimating the pressure loss within the ductwork, fittings, dampers, collectors, and other components. The point where the system curve crosses the fan curve is termed the operating point for the fan. It is at this pressure and cfm that the fan will function with optimum efficiency.

All fan characteristic curves follow the basic relationships given below, and all changes occur simultaneously with a change in fan speed.

(1) The air delivery varies directly with speed. Thus

new CFM = (old CFM)(new RPM/old RPM)

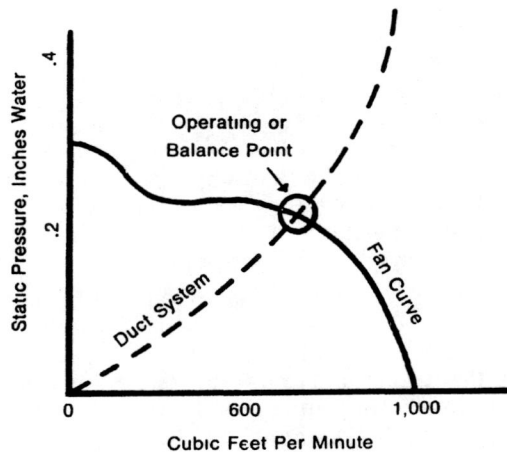

Fig. 5-5. Blower performance curve.

(2) The resistance pressure varies as the square of the speed ratio. Hence,

new SP = (old SP)(new RPM/old RPM)^2

(3) The brake horsepower varies as the cube of the speed ratio, and therefore,

new BHP = (old BHP)(new RPM/old RPM)^3

For example, a fan operating at 600 rpm, delivering 1000 cfm at 0.25 in. WG external static pressure and requiring a brake horsepower (BHP) of 0.06, would have the following characteristics if the fan speed was increased to 1200 RPM:

new CFM = 1000 (1200/600) = 2000
new SP = 0.25 (1200/600)^2 = 1.0
new BHP = 0.06 (1200/600)^3 = 0.48

The proper application of a fan to a system requires not only the basic information about fans but, also, knowledge of the system and the principles of fan-system balance.

The resistance pressure of any duct system can be calculated for any given flow rate. Standard references, such as the ASHRAE Handbook, give complete details on recommended procedures. It is necessary to determine the resistance at only one flow rate, since other points on the system resistance curve can be calculated from the approximate relationship that the resistance pressure (R) varies as the square of the flow rate. Thus, mathematically:

new R = old R (new CFM/old CFM)^2

If, for example, the calculated resistance of a duct system was 0.6 in. WG at 1000 cfm, then at 2000 cfm the new resistance would be calculated as:

new R = 0.6 (2000/1000)^2 = 2.4 in. WG

Because the resistance pressure characteristic line for a fan decreases with the rising air flow rate, and the resistance of a duct system increases with a rise in the air flow rate, the two curves will always cross at one, and only at one, point. This point is the only possible point common to both the fan and the system. It is, therefore, the fan system *balance point*.

Figure 5-6 shows the balance point for a fan and a system. If the fan speed is increased, a new balance point A' will be established at the point where the system curve intersects the fan curve, now at the higher speed. Note that the speed change in no way changes the system curve. The system curve can be changed by (1) dirt on a filter, (2) a damper adjustment, or (3) a basic alteration.

For example, if a new cooling coil were added to the original system, the system resistance at a given flow should be higher. Suppose, that the air flow rate represented by point A, the system resistance with the new coil, rises to that represented by point C. Knowing that resistance varies as the square of the flow rate, a new system resistance curve can be calculated and drawn through point C. This new system curve intersects the two fan speed curves at points B and B' which are the new balance points at the two fan speeds.

From an analysis such as this, a solar system's performance can be predicted before construction or before modifications—often providing valuable money and time saving guides for the dealer-contractor and for the consumer.

Air System Heat Exchanger

One of the components in the duct of the air system is a heat exchanger for warming domestic hot water. It is placed in the basement duct near the hot water tank. Heat exchanger designs may vary, but Figure 5-7 shows a typical unit.

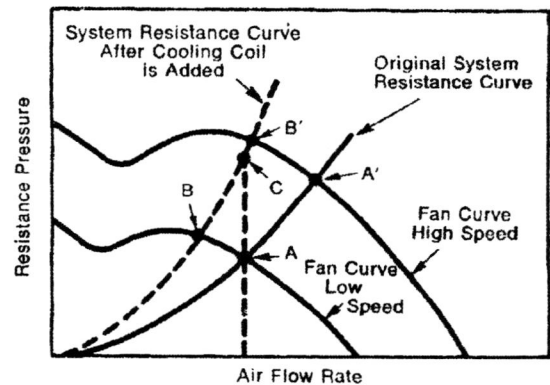

Fig. 5-6. Fan and system balancing.

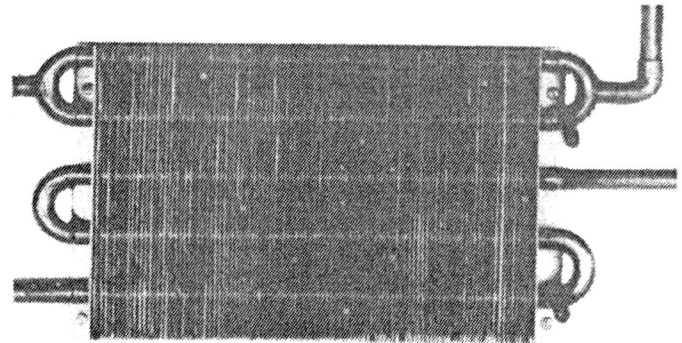

Fig. 5-7. Air to water heat exchanger.

CONTROLS

A control *system* consists of two essential elements. One is a *controller* and the other is a *controlled device* or *actuator*. An example of a controller is an ordinary room thermostat. An example of an actuator or controlled device is a solenoid valve that opens or closes upon command from the thermostat to start or stop the flow of a liquid.

With few exceptions, the heating and air-conditioning technician is familiar with the types of control devices used in solar heating, since they are also used in conventional heating systems as well. As technology develops there will probably be more packaged approaches to solar controls to reduce field installation time. The following are the most commonly used controls.

Two-State Low Voltage Thermostat

A two-stage low voltage thermostat senses room air temperature. The first stage calls for heat directly from the collector or from storage. The second stage turns on conventional furnace, boiler or heat pump. (See Figure 5-8) Single stage thermostats can also be used in conjunction with other devices to perform the same functions. (See Figure 5-8).

(Honeywell)

Fig. 5-8. Two-stage heating, single-stage cooling thermostat.

Outdoor Thermostat

An outdoor thermostat is used to sense outdoor temperature. As in the case of a heat pump application, an outdoor thermostat may prevent auxiliary heat from operating until a preset outdoor temperature is reached. (See Figure 5-9) Fan speed is also sometimes varied using outdoor temperature.

Differential Controller

An ordinary temperature controller is a thermostat that senses water or air temperature in various parts of the system to initiate control action; for example, stop or start fans and pumps and open or close dampers valves. A *differential* temperature controller measures the difference between air or water temperatures at two or more locations in the system as shown in Figure 5-10.

There are any number of specialized differential controllers available to perform multi-control functions in a solar system. For example, one "proportional" differential controller varies pump speed as a function of the temperature difference between collector and storage tank. With only a 3°F difference between collector fluid temperature and storage temperature, the pump operates at slow speed. As the differential rises to 11°F, then the controller switches the pump to high speed.

These specialized differential controllers can also be "wired" to provide high temperature limit for storage and to provide a low temperature turn-on circuit for freeze protection. The pump is started when outdoor temperature is about 37°F.)

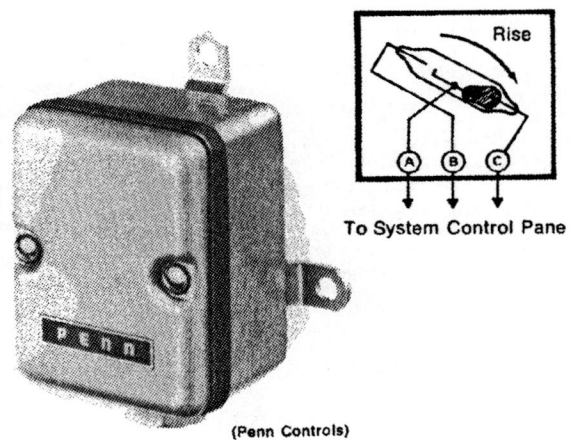

(Penn Controls)

Fig. 5-9. Outdoor thermostat.

5-5

Outdoor Reset Controller

An outdoor reset controller senses outdoor air temperature. It is used to reset the control point of a temperature controller (either up or down) relative to changes in outdoor temperatures. (See Figure 5-11).

Sensors

Every controller is made up of two basic components: a *sensing* device that measures temperature or pressure change, and a *tranducer* that converts that detection into electrical or mechanical action. Common sensors include the simple bimetal, remote bulb, and more recently, thermistor.

Bimetal Sensor. A bimetal sensor consists of two dissimilar metal strips bonded together (copper and iron are examples). They expand at different rates per degree rise in temperature causing the metal strips to bend as illustrated in Figure 5-12.

Bellows/Diaphram Sensor. This type of sensor uses a bulb to sense the temperature. The liquid in the bulb expands in the bellows or against the diaphram and causes the rod to move which contacts the controller mechanism. An example of each type sensor is shown in Figure 5-13.

Thermistor. A thermistor is a semi-conductor material with electrical resistance characteristics that vary with temperature. When connected to an electrical circuit, current flow varies in proportion to changes in the resistance of a thermistor exposed to changing air or liquid temperatures. Through the use of solid state electronics, the current change is amplified to do work—open or close a relay, operate a solenoid, etc. Thermistor sensors can be located as far as 200 feet from the controller. Because of this feature, they are widely used to monitor solar systems. Ordinary 18 and 14 gauge wire can be used to connect sensors to a controller.

Thermistor sensors are relatively small components. They contain a nickel wire-wound element. The ability of a sensor to react to temperature change is called the *temperature coefficient.* For example, a sensor with a temperature coefficient of 3 ohms per degree Farenheit will increase or decrease its resistance, hence, current carrying capability, as ambient temperatures rise and fall. This change could signal the need for heat. The controller to which the sensor was connected would then activate the necessary heating system components.

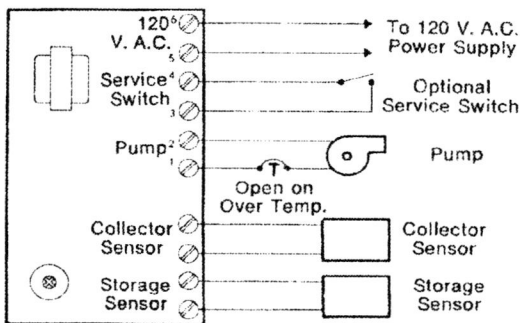

Typical Wiring Diagram

(Penn Controls)

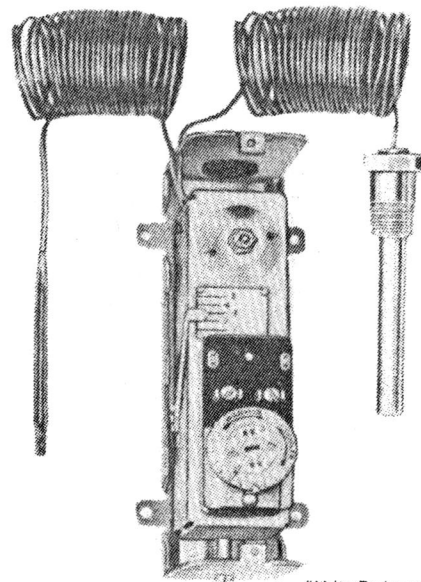

Fig. 5-10. Differential temperature controller.

(White Rodgers)

Fig. 5-11. Reset control.

The sensor in Figure 5-14 is encased in epoxy. It could be installed in an air collector circuit duct at the outlet of the collector to sense the stagnation temperature at which the circuit should be activated to prevent overheating the collector. It can also be placed in the heat storage unit at a location to sense the average temperature within the unit. In an air system, it can be inserted near the bottom among the rocks. For a liquid system, it would have to be placed in a bulb well (Figure 5-15) so that the tank can remain water-tight and not be drained if the thermistor requires service, and it will not make direct contact with the liquid. The corrosive effects of the liquid would damage the sensitivity of the sensor. When this sensor model is placed in a bulb, a thermal conducting compound is used to hold the sensor in the bulb so that the sensitivity to temperature change is not changed significantly. This compound is about the consistency of automobile lubricating grease.

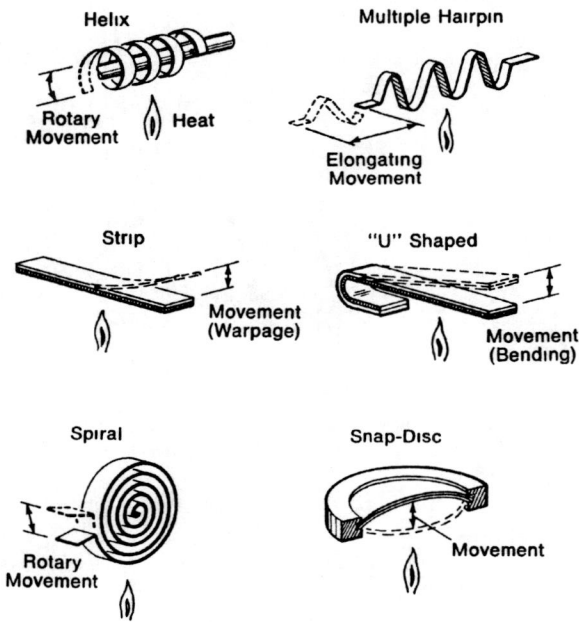

Fig. 5-12. Bimetal sensors.

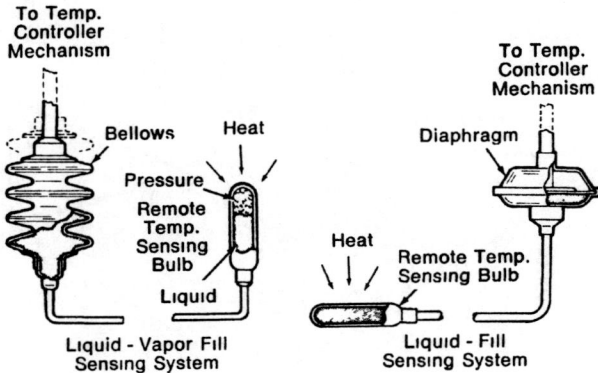

(Penn Controls)

Fig. 5-14. Typical thermistor sensor.

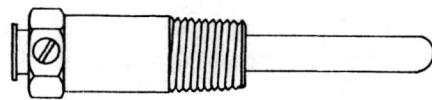

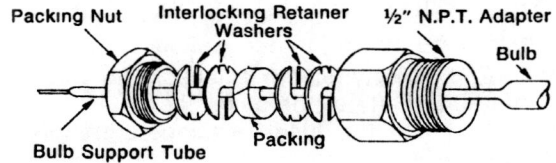

Fig. 5-15. Typical bulb well.

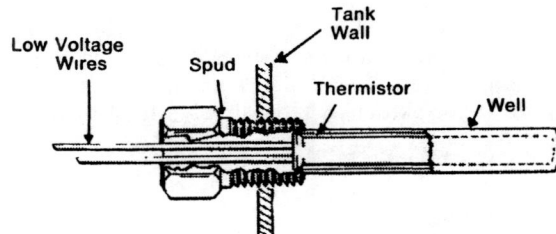

Fig. 5-13. Bellows/diaphragm sensors.

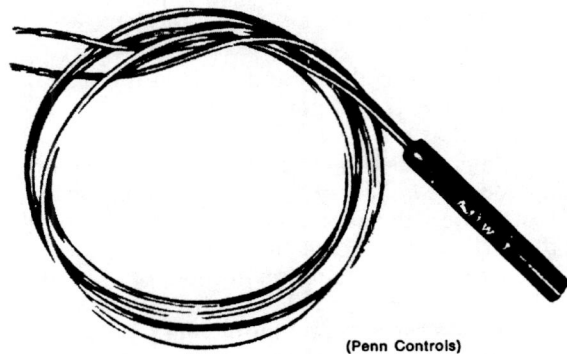

Fig. 5-16. Sensor installed through wall of tank.

There are other bulb well designs which can be used where heat sensing is desirable inside a pipe.

An alternative to using three components—the sensor, the bulb well, and the thermal compound—would be to use a preassembled component as shown in Figure 5-16.

This is a brass unit with the sensor sealed inside and has ½″ pipe threads for mounting in a threaded hole in the heat storage tank or other control locations in the system.

It may be desirable to install a sensor in a piping system without the bulb well. If this is the case, a pipe Tee can be installed in the line and a sensor of the type shown in Figure 5-17 can be used.

The sensor could be mounted in the side or one end of the Tee, depending on the installation specifications.

Placing a sensor near the bottom third of a liquid storage tank may be specified. For this type of installation, the sensor is screwed into a length of pipe. The length of the pipe is predetermined so that the sensor will function at the proper level in the liquid. The sensor screws into the end of the pipe with the wires extending up the pipe and out of the tank. The pipe is secured to the tank with a bushing as shown in Figure 5-18.

Another application of the sensor is shown in Figure 5-19. The sensor is mounted in the sheet metal duct and held in place with sheet metal screws in an air system.

The sensor in Figure 5-20 is the same sensor as in Figure 5-14. However, it is in a more durable copper housing with a hole at the end to mount it more securely to a collector plate or elsewhere. A hose clamp can be used to secure this model on a pipe before the insulation is installed.

Figure 5-21 is typical of a sensor that would be mounted directly to the absorber plate in the collector to sense the absorber temperature indicating availability of solar energy at the collector. The sensor is attached with a sheet metal screw.

Valve

Electrically powered solenoid valves are used to start and stop as well as divert fluid flow in a liquid system. They can be controlled by one or more thermostats. A typical valve is shown in Figure 5-22.

Transformer

A transformer converts live voltage (usually 120V) to a lower voltage (typically 24 volts). This voltage drop is necessary because thermostats

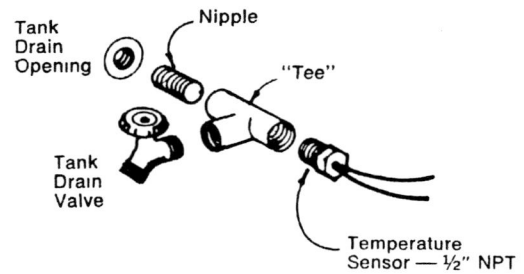

Fig. 5-17. Threaded tank sensor.

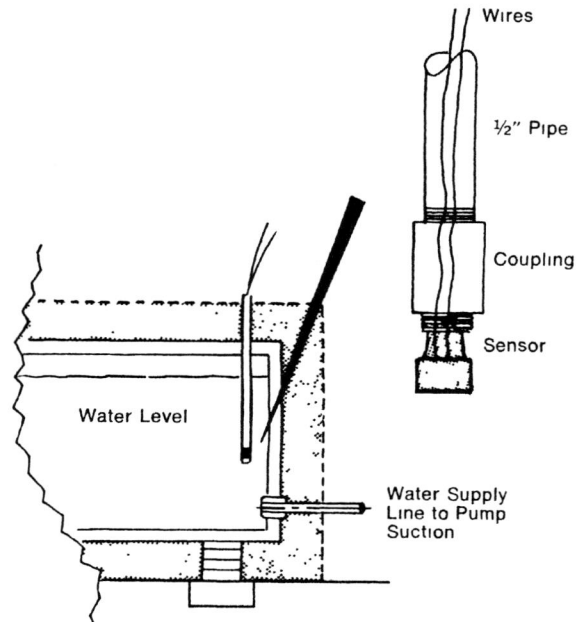

Fig. 5-18. Sensor fastened to end of pipe probe for sensing temperature in deep tanks.

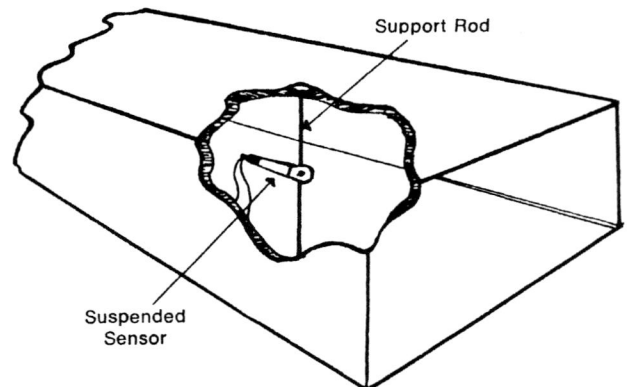

Fig. 5-19. Duct mounted sensor.

and many other control devices are designed for low voltage operation. A typical transformer and its schematic are shown in Figure 5-24.

Damper

Electrically powered multi-blade dampers are used to start and stop, as well as divert, fluid in an air system. They may be thermostatically controlled or opened and closed by the force of air from a blower. Figure 5-23 is an example of a motor-driven damper.

Relay

A relay is an electromagnetic switch. This device (Figure 5-25) will often be operated by some low voltage thermostat (24V) or other controller. When the relay is activated, it can cause a higher voltage load such as a 120 or 240V fan or pump to operate.

A typical transformer-controller-relay-load control circuit is illustrated in Figure 5-26.

Some combination of the control devices explained above is necessary for the safe and automatic operation of a solar assisted heating system. However, there are some specialty items that are not electrically operated.

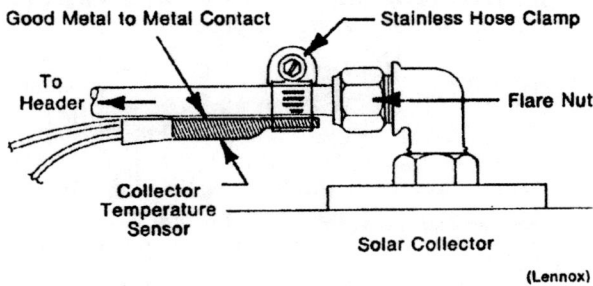

Fig. 5-20. Sensor connected to surface of copper discharge line.

Fig. 5-22. Motorized valve.

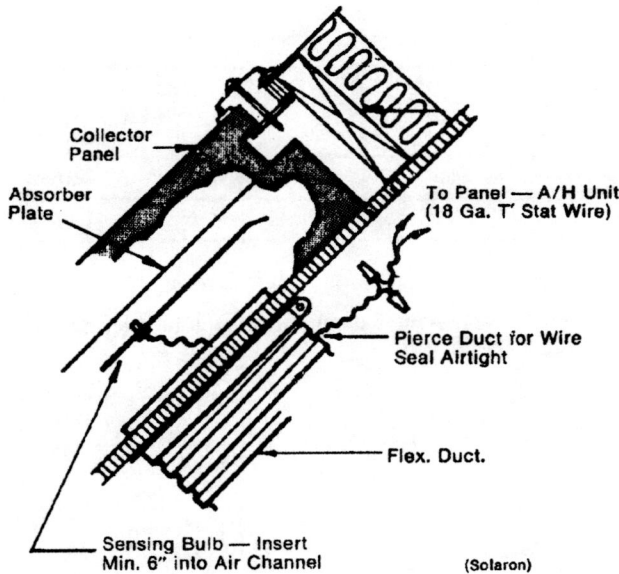

Fig. 5-21. Sensor mounted in air collector.

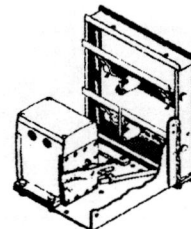

Fig. 5-23. Motorized damper.

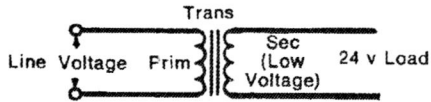

Fig. 5-24. 24 volt transformer.

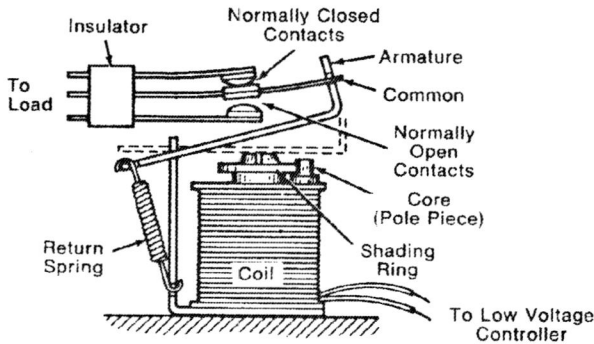

Fig. 5-25. Relay details.

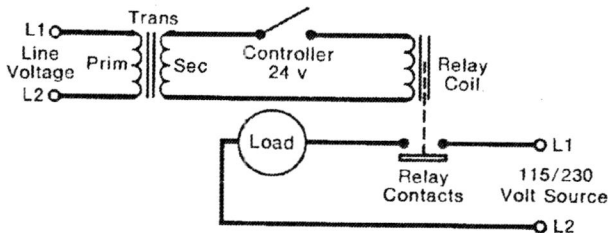

Fig. 5-26. Transformer/relay schematic.

HYDRONIC SPECIALTY ITEMS

Hydronic specialties are non-electrical components that serve some control or system maintenance function. The following devices are representative of specialized components essential to the solar heating system.

Expansion Tank

In completely closed liquid systems, an expansion tank or "air cushion tank" is required to provide "room" for the expansion of heated water. For example, 50 gallons of water at 60° F, when heated to 200° F, would become 51.65 gallons at the elevated temperature. The cushion tank permits this expansion while also controlling the pressure in the system. Open or vented systems usually provide for the expansion within the storage tank as illustrated in Figure 5-27.

Air Vent

An air vent is placed at the highest point in the system (above the collector array). The devices illustrated in Figure 5-28 are of the float type, hydroscoptic disc, and manual design. They vent air from the system as it is being filled with liquid.

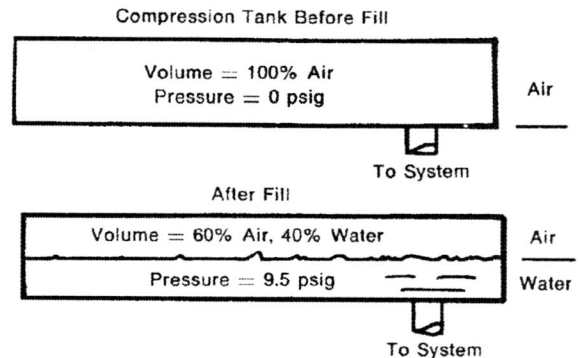

Fig. 5-27. Conventional expansion tank (above) diaphragm design (below).

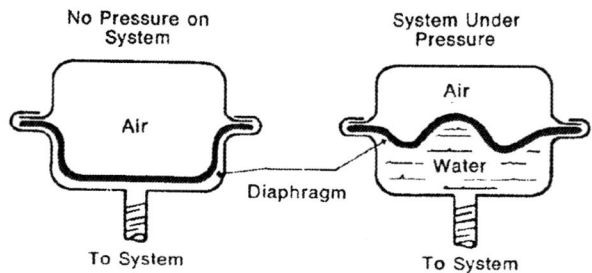

Check Valve

This type of valve limits fluid flow to one direction. There are a number of designs available for horizontal and vertical piping installation as illustrated in Figure 5-29. A flow control valve is a specialized check valve which prevents gravity circulation in the solar circuits when the pumps are off.

Vacuum Breaker

A vacuum breaker is sometimes used to relieve an unwanted vacuum condition in a drain down system. This valve, shown below in Figure 5-30, permits the system to drain efficiently by gravity, by admitting atmospheric pressure into the return piping.

Pressure Relief Valve

A pressure relief valve, also known as a safety relief valve, is pressure operated to prevent excessive pressure in a closed system in the event of any malfunction. Figure 5-31 depicts one of these specialty items.

Balancing Valve

A balancing valve is a simple, inexpensive, square head cock. It is used to adjust flow rate in each circuit and especially through the collector array (see Figure 5-32). There are special, more deluxe, balancing valves used, as can be seen in Figure 5-33.

Pressure Reducing Valve

A pressure reducing valve is often used as part of the water supply system. Water service pressure that is too high for some of the heating system components can be reduced by using a valve such as illustrated in Figure 5-34.

Air Eliminator

An air eliminator (Figure 5-35) is used in conjunction with an air cushion pressure tank to help purge a closed system of unwanted air.

Fig. 5-29. Swing check valve —one of several designs.

(Johnson Corp.)

Manual Vent

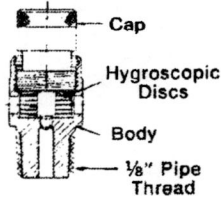

Cap

Hygroscopic Discs

Body

⅛" Pipe Thread

Automatic Vent

Fig. 5-28. AUTOMATIC air vent (above) operates by means of hygroscopic discs that when wet expand and seal venting ports. If air accumulates, discs dry and contract opening ports and venting air. (Dunham-Bush, Inc.) Float type vent (below) operates by means of a float that opens and closes a valve. In (A) water level is high and vent is closed. In (B) air has accumulated and water level is low causing float to open vent. (Armstrong Machine Works.) Manual vent (above) must be opened and closed by hand. (ITT Bell & Gossett.)

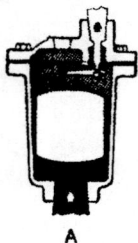

Automatic Vent

A   B

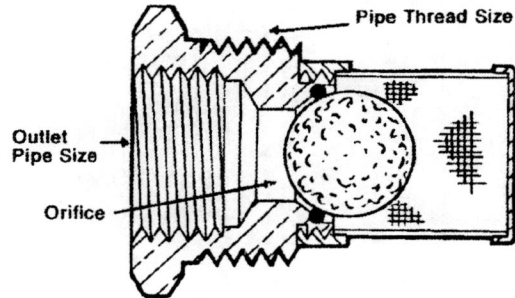

Pipe Thread Size

Outlet Pipe Size

Orifice

Fig. 5-30. Vacuum breaker.

Dielectric Union

Several dielectric unions may be needed in the system. Connecting fittings made of certain metals, as discussed in Lesson Four, may result in corrosion and restriction of liquid flow throughout the piping. Corrosion is most apt to occur when ferrous (iron) and non-ferrous (aluminum, brass, copper) fittings and components are connected together. One way to eliminate the problem is with plastic fittings or to use dielectric unions as shown in Figure 5-36.

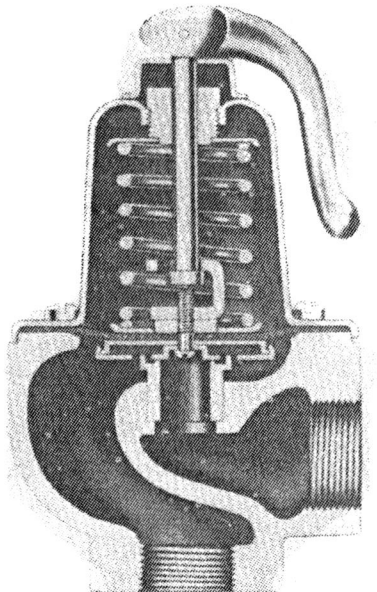

(Thrush)

Fig. 5-34. Pressure reducing valve.

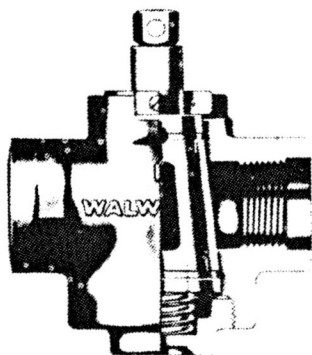

(ITT Bell & Gossett)

Fig. 5-31. Pressure relief valve.

(ITT Bell & Gossett)

Fig. 5-33. Calibrated balance valve.

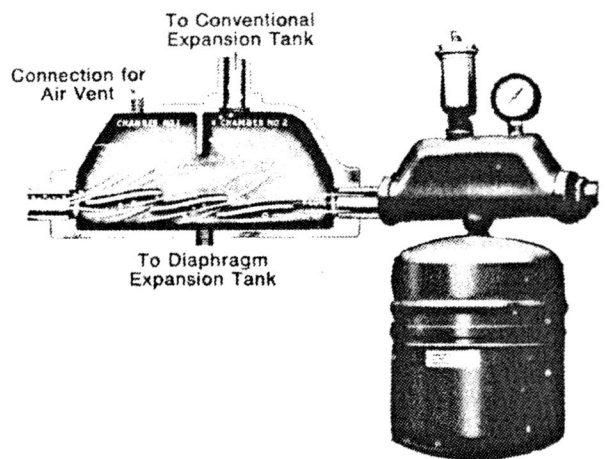

(Walworth Co.)

Fig. 5-32. Section view of plug or square head cock used to adjust flow in multiple circuit systems

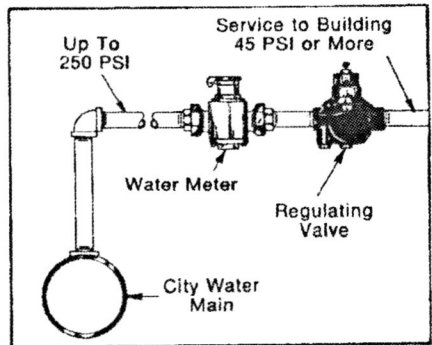

(Taco)

Fig. 5-35. Special air elimination device connected to diaphragm tank.

5-12

Fill System

Filling the collector loop of a solar assisted heating system may be done automatically or manually. Manual filling practices must be followed if local building codes prohibit connection of a solar system containing a toxic fluid (antifreeze) to city water service. An automatic fill system may be a convenience to the occupant but the (glycol) antifreeze may become diluted over a period of time. This would create a maintenance problem if freezing occurred. The manual fill system would include a pressure reducing valve, a check valve, and a manually operated globe valve.

An automatic system filling valve (Figure 5-37) has a built-in strainer and check valve mechanism.

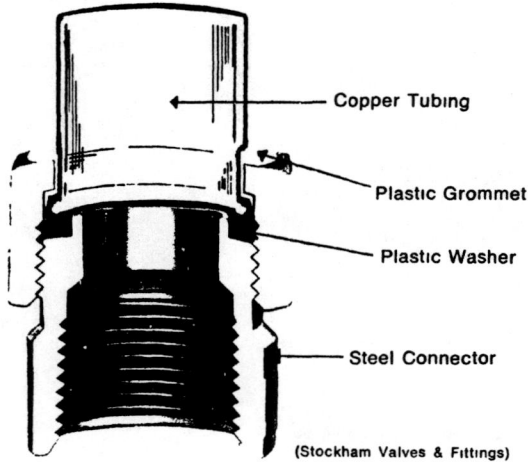

Copper Tubing

Plastic Grommet

Plastic Washer

Steel Connector

(Stockham Valves & Fittings)

Fig. 5-36. Dielectric union.

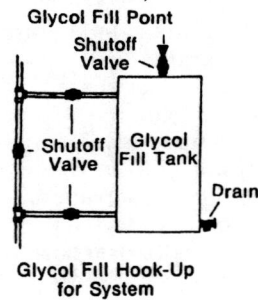

Glycol Fill Point

Shutoff Valve

Shutoff Valve

Glycol Fill Tank

Drain

Glycol Fill Hook-Up for System

Fig. 5-38. Glycol fill system.

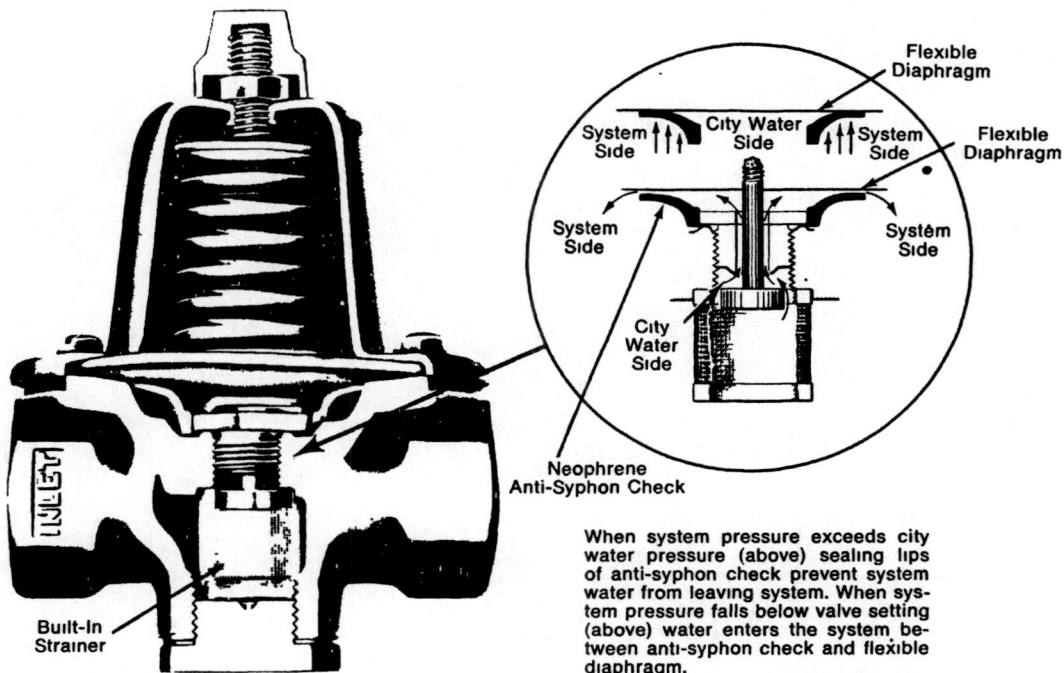

Flexible Diaphragm

Flexible Diaphragm

System Side

City Water Side

System Side

System Side

System Side

City Water Side

Built-In Strainer

Neophrene Anti-Syphon Check

When system pressure exceeds city water pressure (above) sealing lips of anti-syphon check prevent system water from leaving system. When system pressure falls below valve setting (above) water enters the system between anti-syphon check and flexible diaphragm.

(ITT Bell & Gossett)

Fig. 5-37. Automatic fill valve with anti-syphon feature.

Antifreeze-Filling Hook-Up

For a closed loop system that contains anti-freeze, a fill system for the glycol must be included. One example of such a system is in Figure 5-38.

Summary

Many kinds of controls and specialty items are needed for a solar assisted heating system. Centrifugal pumps are used for liquid systems. Fans are used for air systems. When water heating for occupant use is included, the systems must include heat exchangers. Numerous electrical and/or heat activated control devices assist in effective operation of the system. Some non-electrical specialty items are also required. Shut-off valves and unions are also needed to isolate major components in a liquid system so they can be removed for repair. In addition, various hose bibs for drain and fill service are used.

All of the control devices and specialty items must be made safe before the system is operated. This means that all electrical wiring and plumbing installations must be made according to various local codes and licensing practices.

SIZING SOLAR SYSTEM COMPONENTS 6

The basic components of a solar heating system have been reviewed in Lessons Three, Four, and Five. The purpose of this lesson is to provide an understanding of how to select the appropriate size for each of the solar components in the heating system. Figure 6-1 depicts a liquid-to-air solar heating system. Ten basic components of a typical installation are identified in the illustration.

Sizing Solar Collectors

The initial step in sizing the various components in a solar heating system is to *size* the collector. Because collector Btu output and system utilization of these Btu's is dependent upon a number of weather and component variables, a great deal of research has been conducted to study system performance and to develop practical collector sizing procedures. Presently, there are several sizing procedures proposed and used throughout the industry. These vary from simple rules of thumb, such as the number of square feet of collector as a fraction of the square feet of the home (e.g., collector area equals from ⅓ to ½ the floor area), to highly sophisticated computer programs. Many collector manufacturers have selected a particular procedure and refined it for use with their respective proprietary collectors.

Tables for Selecting Collector Size

For example, Table 6-1 shows a tabulation of the output of a specific manufacturer's collector panel in thousands of Btu's for each month of the winter heating season for selected cities in the United States. The table indicates that in January, for the city of Columbus, Ohio, a single collector panel from this manufacturer will supply 280,200 Btu's. Armed with this information and the monthly energy needs of the structure based on building heat loss and monthly degree days, the number of panels required to supply some fixed percentage of the heating requirements can be determined.

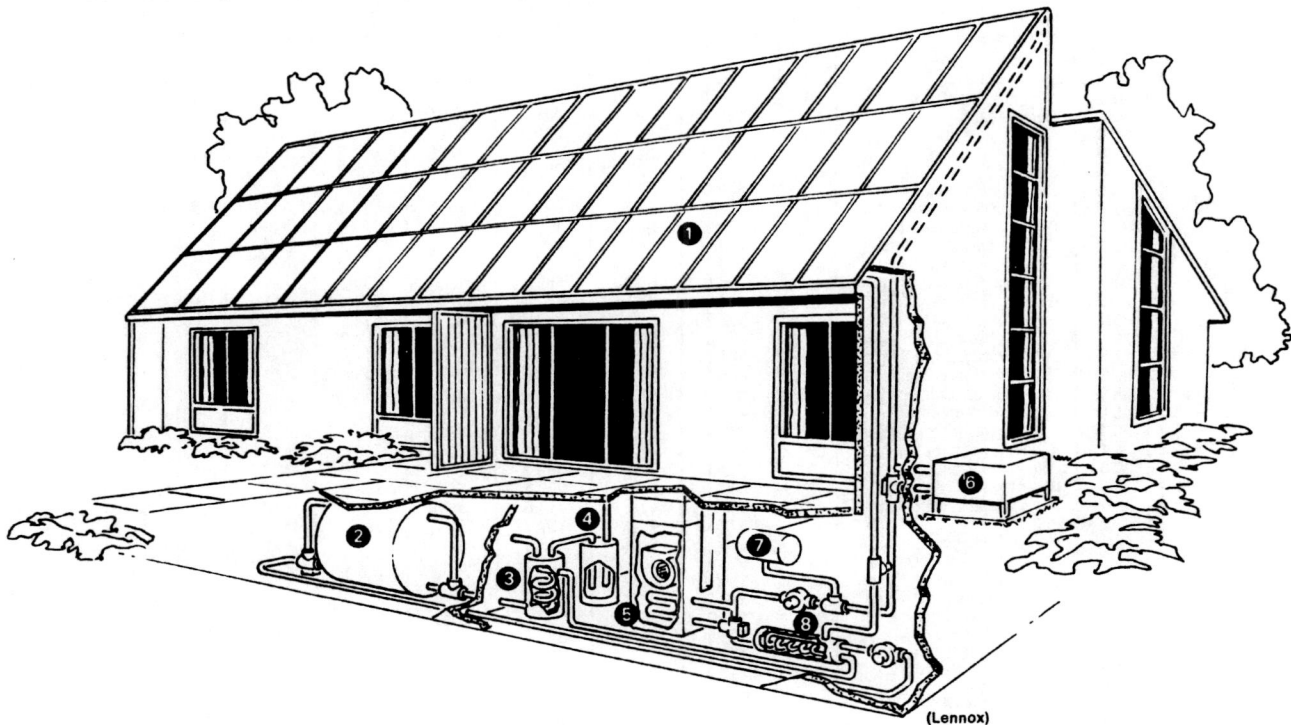

(Lennox)

Fig. 6-1. Key components in a liquid to air solar system: 1. Solar collectors, 2. Storage tank, 3. Hot water heat exchanger, 4. Hot water holding tank, 5. Space heating coil, 6. Purge coil (releases excess solar heat), 7. Expansion tank, 8. Heat exchanger.

Another example of a simplified procedure is shown in Table 6-2. This table gives a simple divisor termed "LC" for 85 cities that can be divided into a building's heat loss *per degree day* to arrive at a collector size to supply a specific fraction of the heating load. This is based on simulation studies made at Los Alamos Scientific Laboratory. Here is how it works.

Suppose a house located in Atlanta, Georgia is to be solar heated. Assume the heat loss is 40,000 Btuh for 20° F average outside and 70° F average inside temperature. Table 6-2 illustrates how the calculation is made.

A third example of a simplifed procedure can be found in Table 2-1 of the SMACNA Installation Standards. Figure 6-3 shows a portion of this table.

In this table, a separate divisor is provided for air and liquid collector systems. Further, there are two choices of collector tilt and a selection for 30, 50 or 70 percent solar contribution. Also, in this case, the divisor is simply divided into the calcu-

lated design heat loss which differs from the previous example. Here is how to use the SMACNA table.

Consider the previous example city, Atlanta, Georgia, and the building with a 40,000 Btuh heat loss. For a liquid system, Table 2-1 provides divisors of 316 to obtain a 30 percent solar contribution, 152 for 50 percent and 84 for a 70 percent contribution for a liquid collector tilted at 53 degrees. Based on a 50 percent contribution:

$$\frac{40,000}{152} = 263 \text{ square feet of collector}$$

Note, first, that the estimated collector area determined from the procedures used in Figure 6-4 and above differ substantially (325 vs 263). Second, the Los Alamos procedure made no distinction between air or liquid systems. Also, the designer had a choice of tilt angles in the SMACNA procedure but he did not in the Los Alamos approach. (Los Alamos assumed a tilt

Table 6-1. Btu output per panel per month — one way to simplify sizing.

MONTHLY OUTPUT,/COLLECTOR PANEL — MBTU									
°Latt. North	Location	Collector Tilt	October	November	December	January	February	March	April
48 2	Glasgow, Montana	60°	496 2	338 8	264 3	333 6	436 7	536 7	450.5
43 6	Boise, Idaho	55°	543 2	387 4	288 3	313.1	404 4	494.4	481 9
40 0	Columbus, Ohio	55°	420 0	257 9	259 3	280.2	312.1	394.0	360 8
35 4	Oklahoma City, Oklahoma	50°	637 3	554.2	508 1	504 6	494 2	556.3	481.9
40	Salt Lake City, Utah	55°	536 6	407 4	392 2	345 4	406 6	475 3	440.4
29 5	San Antonio, Texas	40°	660 6	537.6	497.3	527 5	529 7	590 8	484.4
32 8	Fort Worth, Texas	45°	668.8	579 8	498.6	488.8	497.4	598.6	511.6
40 3	Grand Lake, Colorado	55°	509 9	386.4	370 4	390.8	450.9	410.5	415 0
42.4	Boston, Massachusetts	55°	425 4	307 6	279.1	303 0	319 2	389.7	339 8
27.9	Tampa, Florida	40°	660.0	656 6	610.0	646 7	608.0	670 0	578.0
33 4	Phoenix, Arizona	45°	777.0	663.0	589 7	606.2	655 7	756.0	678.0
33.7	Atlanta, Georgia	45°	566.3	489.5	422.7	423.8	441 7	518.0	499 0
35.1	Albuquerque, New Mexico	50°	719 2	628 7	580 8	604.9	592 9	682.3	588 0
40 8	State College, Pennsylvania	58°	468 6	307.5	255.8	280.6	312.9	405.8	375.5
42 8	Schenectady, New York	55°	381 8	242.1	315 1	281.7	317.9	365.2	319 5
43 1	Madison, Wisconsin	55°	465 5	306 7	293 9	321.0	343 5	442 6	370.2
33 9	Los Angeles, California	50°	604.3	577 2	540 4	535 4	556 8	633 0	482.9
45 6	St Cloud, Minnesota	60°	419 8	309 3	274 4	362.7	416.3	482.6	381.6
36.1	Greensboro, North Carolina	50°	537 8	465 1	396 0	409.5	430 2	481.4	456 4
36.1	Nashville, Tennessee	50°	556.0	424.0	354 5	325 7	380 7	456 8	438 2
39 0	Columbia, Missouri	50°	559 1	438 7	339 3	356.7	390.9	485.7	440 9
30 0	New Orleans, Louisiana	40°	589.7	506 8	390 5	415.8	396 6	473 5	448.2
32.5	Shreveport, Louisiana	45°	590 4	475 3	419.6	459 6	452 0	534.7	461.1
42.0	Ames, Iowa	55°	378 4	262 9	197.7	238 2	294 9	368.2	354.7
42.4	Medford, Oregon	55°	455 1	298 4	213 4	255.8	343.4	434.1	412.3
44.2	Rapid City, South Dakota	60°	550 9	436 3	383 3	405.1	453.2	518.0	420.3
38.6	Davis, California	50°	719.3	536 9	401.3	448.9	515 6	666.1	647.2
38 0	Lexington, Kentucky	50°	616 5	477.4	377 8	359.6	411 1	487 5	479.9
42.7	East Lansing, Michigan	55°	415.0	261.7	230.1	247.5	314 2	387 4	313 4
40.5	New York, New York	55°	505 8	389.4	328 0	357.1	396 6	451 9	381.8
41.7	Lemont, Illinois	55°	477 1	352 7	321 4	343.6	373 9	450 3	365.4
46 8	Bismark, North Dakota	60°	491.1	346.6	279 9	335.6	405 7	476 6	411.9
39.3	Ely, Nevada	55°	636 7	559.0	475.8	481.1	512 4	597.5	483.4
31.9	Midland, Texas	45°	651 5	596 2	540 7	543 0	543 1	648.9	562 1
34.7	Little Rock, Arkansas	50°	578 4	470 0	400 0	383 6	408 1	486.5	435 4
39 7	Indianapolis, Indiana	55°	501 7	354 3	293 2	300.1	332 1	417 6	360 4

(Fedders)

6-2

angle of latitude plus 10 degrees which, for Atlanta, would mean a collector tilt of 44 degrees).

The important point in this comparison is that "assumed conditions" for specific simplified procedures are not always the same. Before a designer/technician uses any simplified approach for collector sizing, there should be a thorough understanding of the assumptions made by the developers of the design technique. The designer/technician should also be aware of how these assumed parameters differ for specific applications.

FCHART Collector Selection Technique

Perhaps the best known detailed sizing procedure and the one most frequently referred to for specific collectors is termed the FCHART procedure developed at the University of Wisconsin. (Table 2-1 in the SMACNA Standards is based on this procedure.) Figures 6-4 and 6-5 show the FCHART for liquid systems and air systems respectively. Each "f" curve represents the fraction of the monthly energy demand supplied by solar energy as a function of the "X" and "Y" coordinates. The horizontal scale (X) is the ratio of the monthly solar collector losses to the monthly heating demand; the "Y" scale is the ratio of the monthly solar energy absorbed by the collector to the monthly heating demand. Since values of X and Y vary each month, the fraction (f) of the heating demand satisfied by the collector will also vary each month.

The curves were derived from computer simulation studies of two "standard" air and liquid solar heating systems. To use the FCHART, the slope and intercept of the efficiency curve for the specific manufacturer's collector being used must be known (see Lesson 3, Figure 3-12, on collector efficiency). In addition, the designer must know the monthly incident solar radiation, monthly degrees days, building heat loss, and

Table 6-2. "Divisors" to determine collector area required to supply a fixed percentage of energy.

City, State	Latitude (°N)	Elevation (ft)	Degree-days	LC (Btu/degree-day-ft°) where solar provides 25%, 50%, 75% of total heat		
				25%	50%	75%
Los Alamos, NM	36	7200	6600	107	41	21
Columbus, OH	40	760	5211	77	29	13
Corvallis, OR	45	236	4726	120	42	18
Davis, CA	39	50	2502	198	72	33
East Lansing, MI	43	878	6909	76	28	13
East Wareham, MA	42	50	5891	97	37	18
El Centro, CA	33	12	1458	547	206	97
Flaming Gorge, UT	41	6273	6929	111	43	21
Granby, CO	40	8340	5524	119	47	23
Toronto, Canada	44	443	6827	72	27	13
Griffin, GA	33	1001	2136	217	84	42
Winnipeg, Canada	50	820	10629	63	23	11
Ithaca, NY	42	951	6914	68	24	11
Inyokern, CA	36	2186	3528	232	88	42
ANL, Lemont, IL	42	750	6155	79	30	14
Newport, RI	41	50	5804	97	37	18
Laramie, WY	41	7240	7381	106	42	21
Page, AZ	37	4280	6632	128	48	23
Prosser, WA	46	840	4805	117	41	18
Pullman, WA	47	2583	5542	100	36	16
Put-In-Bay, OH	42	580	5796	68	24	11
Richland, WA	47	731	5941	100	35	15
Raleigh, NC	36	440	3393	133	52	25
Riverside, CA	34	1050	1803	391	152	74
Seattle, WA	48	110	4785	94	33	13
Sayville, NY	41	56	4811	98	38	18
Schenectady, NY	43	490	6650	63	24	11
Seabrook, NJ	39	110	4812	97	37	18
Shreveport, LA	32	220	2184	179	70	35
State College, PA	41	1230	5934	78	29	14
Stillwater, OK	36	910	3725	132	52	25
Tallahassee, FL	30	64	1485	288	113	57
Tucson, AZ	32	2440	1880	301	118	59
Oak Ridge, TN	36	940	3817	111	42	20
Fort Worth, TX	33	574	2405	185	73	37
Lake Charles, LA	30	60	1459	244	96	48
Apalachicola, FL	30	46	1308	324	129	65
Brownsville, TX	26	48	600	517	218	110
San Antonio, TX	30	818	1546	262	103	52
Greensboro, NC	36	914	3805	128	50	24
Hatteras, NC	35	27	2612	204	79	39
Atlanta, GA	34	1018	2961	154	59	29
Charleston, SC	33	69	2033	210	82	41
Nashville, TN	36	614	3578	117	44	21
Lake Charles, LA	30	39	1459	261	104	53
Little Rock, AR	35	276	3219	126	48	24
Oklahoma City, OK	35	1317	3725	134	53	26
Columbia, MO	39	814	5046	102	38	18
Dodge City, KA	38	2625	4986	126	49	24
Caribou, ME	47	640	9767	68	26	12
Burlington, VT	44	385	8269	63	24	11
Blue Hill, MA	42	670	6368	82	31	15
Cleveland, OH	41	871	6351	71	26	12
Madison, WI	43	889	7863	76	28	13
Sault Ste Marie, MI	46	724	9048	74	27	12
Saint Cloud, MN	46	1062	8879	71	27	13
Lincoln, NE	41	1316	5864	104	39	19
Midland, TX	32	2885	2591	202	79	39
El Paso, TX	32	3954	2700	228	88	44
Albuquerque, NM	35	5327	4348	161	64	31
Grand Junction, CO	39	4832	5641	119	46	22
Ely, NV	39	6279	7733	119	47	23
Las Vegas, NV	36	2188	2709	218	84	42
Phoenix, AZ	33	1139	1765	300	118	59
Reno, NV	39	4400	6632	125	47	22
Santa Maria, CA	35	289	2967	353	142	67
Bismark, ND	47	1677	8851	78	29	14
Lander, WY	43	5574	7870	108	42	21
Glasgow, MT	48	2109	2996	105	41	20
Rapid City, SD	44	3180	7345	97	37	18
Salt Lake City, UT	41	4238	6052	107	40	19
Boise, ID	44	2895	5809	108	39	17
Great Falls, MT	47	3692	7750	93	35	16
Spokane, WA	48	2356	6655	90	31	14
Medford, OR	42	1321	5008	107	38	16
Los Angeles, CA	34	540	2061	416	157	75
Fresno, CA	37	336	2492	195	70	32
Silver Hill, MD	39	292	4224	111	43	21
Cape Hatteras, NC	35	27	4612	189	74	36
Sterling, VA	39	276	4224	111	43	21
Indianapolis, IN	40	819	5699	86	32	15
Astoria, OR	46	22	5186	127	45	19
Boston, MA	42	157	5624	86	33	16
New York, NY	41	187	4871	88	34	16
North Omaha, NE	41	1323	6612	89	34	16

(Los Alamos Scientific Laboratory)

average monthly outdoor temperature. There are also "correction factors" to be applied for "non standard" variations in system configuration, such as larger or smaller storage capacity than "standard," and heat exchanger performance for liquid systems.

Btu/degree day $= \dfrac{\text{Btu/Hr}}{(t_i - t_o)} \times$ 24 hrs/day

t_i = average inside temperature
t_o = average outdoor temperature

$= \dfrac{40,000}{(70-20)} \times$ 24 hrs/day

$=$ 19,200 Btu/degree day

From the table in Figure 6-3 the LC division would be:

154 to provide 25% of the heating demand
59 to provide 50% of the heating demand
29 to provide 75% of the heating demand

Using the 50% contribution, the size of the collector can be calculated as follows

$\dfrac{\text{Btu/degree day}}{\text{LC}}$

substituting from above

$\dfrac{19,200}{59} =$ 325 4 square feet of collectors required.

Fig. 6-2. Calculating the BTU/degree day and solar collector size using LC value.

Values of X and Y are calculated for each month of the year for the locality in question and for an assumed collector area. Monthly values of "f" are obtained from the chart at the intersection of the X and Y coordinates. The monthly solar energy contributions are then totalled for the season and divided by the total heating demand to determine a seasonal value of "f". The procedure is then repeated for several other assumed collector areas.

The results could be as follows for a specific building and collector type; collector area (assumed) of 400 square feet—seasonal "f" equals .43; collector area increased to 600 square

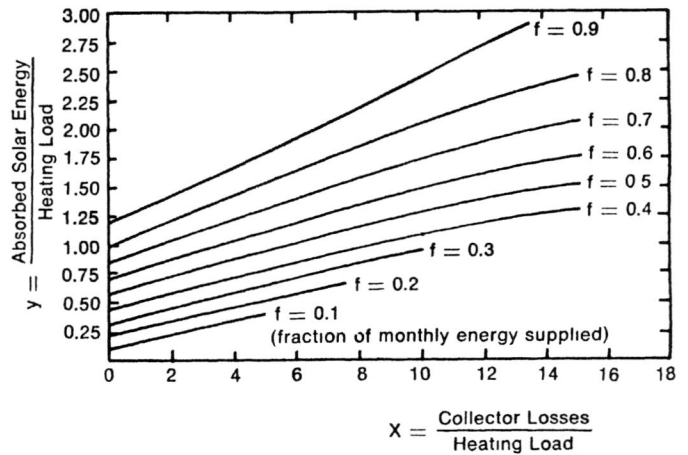

$$x = \frac{\text{Collector Losses}}{\text{Heating Load}}$$

Fig. 6-4. f-Chart for liquid-based solar heating systems.

TABLE 2-1 cont'd
SOLAR CONVERSION FACTORS
UNITED STATES, AUSTRALIA AND CANADA

Location	Design Temp. Difference	AIR						LIQUID					
		Portion of Load Carried by Solar											
		30%		50%		70%		30%		50%		70%	
		37°	53°	37°	53°	37°	53°	37°	53°	37°	53°	37°	53°
Georgia													
Atlanta	55°	292	297	141	150	80	86	310	316	144	152	78	84
Griffin	48°	321	337	162	170	98	105	350	367	166	177	93	101
Idaho													
Boise	60°	196	205	89	99	45	52	205	217	88	99	40	48
Pocatello	72°	235	245	123	123	60	68	252	266	114	125	56	66
Twin Falls	62°	186	189	86	91	44	50	193	197	84	90	40	46
Illinois													
Chicago	70°	74	78	29	32	14	16	67	70	25	28	15	15
Lemont	70°	162	167	77	83	42	47	167	176	74	82	38	43
Indiana													

Fig. 6-3. Portion of Tab. 2-1 SMACNA Installation Standards.

6-4

feet—"f" increases to .55; and for a collector area of 800 square feet—"f" equals .64. Thus, to supply just over half the heating needs (.55), a collector of 600 square feet would be required in this hypothetical case.

The University of Wisconsin has developed a computer program utilizing the FCHART procedure. It is available for sale for private use and a number of collector manufacturers make this service available to their customers. Details of the manual procedure to use FCHART are included in HUD's *Intermediate Minimum Property Standards Supplement.*

Between the highly sophisticated FCHART computer analysis and the drudgery of manual calculations, there is a hand calculator approach devised by researchers at Colorado State University. Termed "Relative Areas Analysis," a designer equipped with special tabulated data for specific cities can simply "plug in" a few numbers into a calculator and determine an annual load fraction. Details of the procedure are contained in a thesis written by C. Dennis Barley, Department of Mechanical Engineering, CSU, Fort Collins, Colorado 80523. The procedure also involves an economic analysis to determine the "best size" of collector. The concept of life-cycle-costing will be discussed next.

ECONOMICAL COLLECTOR SIZING

There are many reasons why a customer/client may choose to purchase a solar heating system. Among them is a concern for the environment, fear of fossil fuel shortages, and a desire to have a new and innovative heating system. However,

when the time comes to pay for the heating system, nearly all customers/clients will want to know how much the solar system will save *before* deciding to purchase. In terms of economic benefit, what size system might be best for a customer; one that contributes 20, 30, 50, or 80 percent of the energy need? Figure 6-6 illustrates a typical economic analysis. The top curve relates collector area versus fraction of the load supplied, as might be determined from FCHART, SMACNA, the LC Table, or other simplified procedures.

The lower curve relates collector area to money saved over a 20 year period. In this example, the peak savings are realized for a collector area of 630 square feet that would result in a 60% solar contribution.

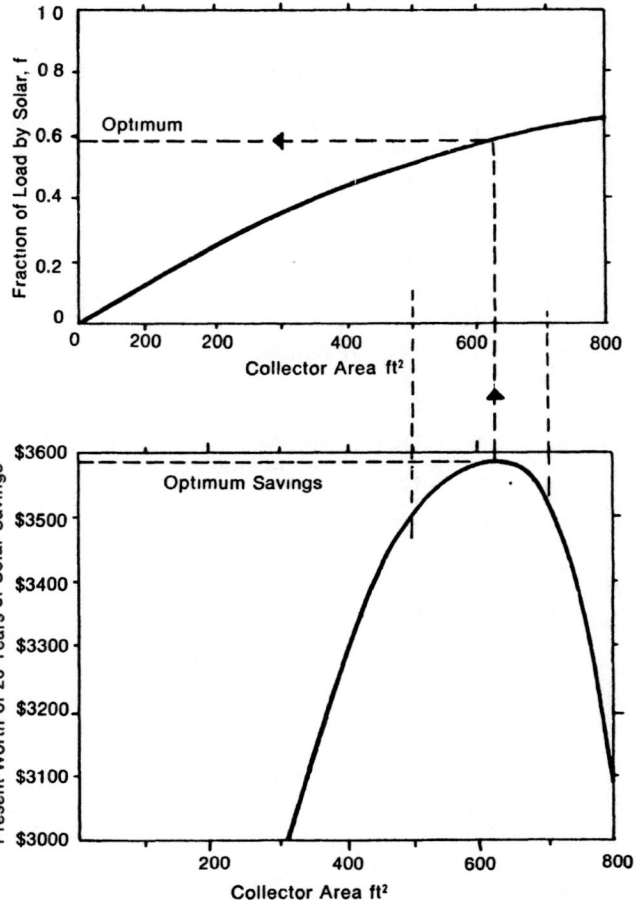

Fig. 6-6. Savings versus collector area.

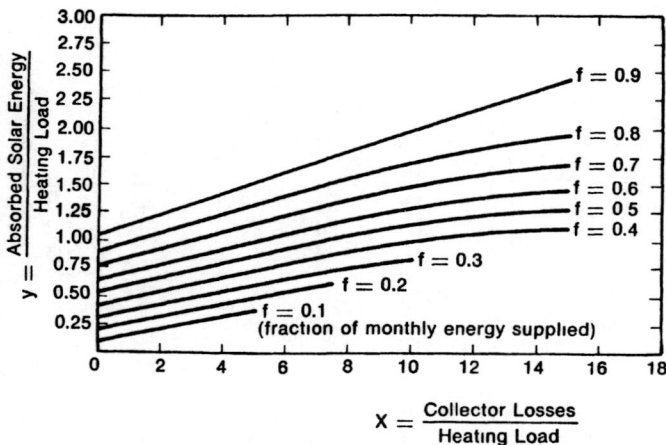

Fig. 6-5. f-Chart for solar air heating systems.

Please note, however, that a modest change of $100 out of the projected $3600 savings in this example would alter the desirable collector size from a low of 500 square feet to perhaps 700 square feet. Thus, the optimum plateau is fairly flat and this means one's choice is rather broad in terms of optimizing payback, especially because of the many assumptions made to complete an economic analysis. Presently, it appears that collectors selected to serve from 50 to 70 percent of the heating load are most economic in normal installations.

LIFE-CYCLE-COSTS

Because of the need to install a combination system (solar plus conventional heating equipment), it is obvious that it will be impossible to create a solar assisted heating system which is less expensive than a conventional system based on initial costs. The sale will be made based on what is termed "life-cycle-cost." (See Figure 6-7.) The customer/client must be convinced that the savings in energy cost over the years the system will actually last (before it wears out) will offset the initial cost of installing the solar heating system. In making the determination of life-cycle-costs, it is necessary to consider the following factors:

1. Solar System Fixed initial cost.

2. Solar collector installed cost per square foot.

3. Loan interest rate.

4. Loan term.

5. Loan down payment.

6. Property tax rate.

7. Income tax rate.

8. Maintenance costs.

9. Insurance.

10. Property taxes.

11. Present fuel costs.

12. Inflation

Obviously, these factors will vary according to the location of the installation. It will be necessary to become familiar with the values for each of the factors for a specific location. The savings versus collector area as shown in Figure 6-6 involves a great deal of manual calculation. Building owner cash flows are calculated for each year of the analysis for both solar and non-solar space

heating installations. By comparing the present values of the yearly costs of the solar and non-solar systems, the economic feasibility of the solar system is determined. Present value refers to the savings in term's of today's dollars. As one knows, the value of money is time-related, and is a normal factor to consider when analyzing investments.

"SOLCOST" COMPUTER ANALYSIS

Because of the numerous calculations involved, the computer can be put to ideal use. An ERDA funded program called "Solcost" is available to contractors to determine optimum collector size. The cost is from $10 to $20 for sizing and life-cycle-costing. For an additional $40, a heat loss calculation can be obtained by computer. Figure 6-8 illustrates the type of Solcost residential analysis that would be provided.

For further information on the Solcost procedure contact: International Business Services, Solar Group, 1010 Vermont Avenue, Washington, D.C. 20005, telephone (202) 628-1450.

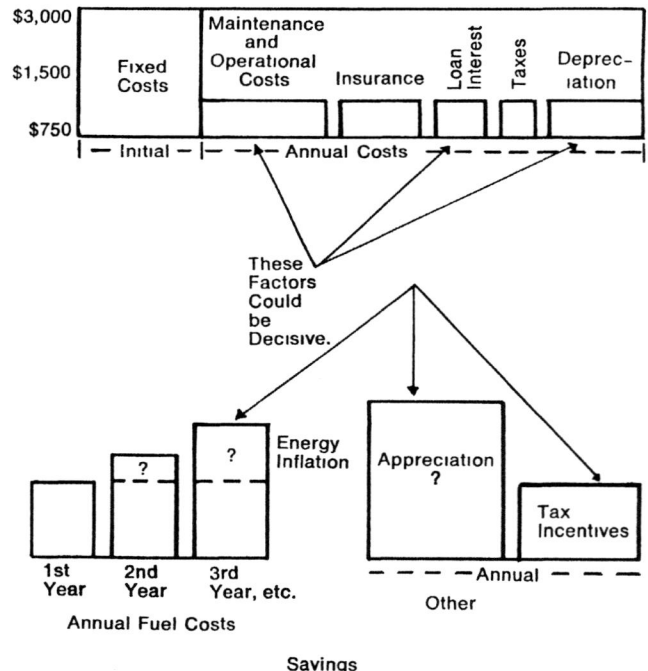

Fig. 6-7. Life cycle costing is a method whereby the total costs of a product can be measured against the annual savings, showing the buyer approximately when his or her investment is paid for.

INPUT

Input Parameter	User Input
Solar System Type	1
Fuel Type for Reference Heating System	2
Fuel Type for Solar Auxiliary Heating System	2
Collector Type	3
Collector Tilt Angle	55 (Degrees)
Collector Azimuth Angle	+ 10. (Degrees)
Site Location	DENVER
Building Heat Loss Coefficient	8 3 (BTU/Sq Ft /Deg.-Day)
Building Floor Area	1950 (Sq. Feet)
Solar System Fixed Initial Cost	$1000.
Solar Collector Installed Cost/Sq. Ft.	$12 00
Loan Interest Rate	09 (9 percent)
Loan Term	20. (Year)
Loan Down Payment	22 (22 percent)
Property Tax Rate	02 (2 percent)
Income Tax Rate	30 (30 percent)
Inflation of Maint , Insur. Property Taxes	04 (4 percent)
Present Electricity Cost $/Kw-hr	$.035
Electricity Cost Escalation Per Year	10 (10 percent)

EXPLANATION OF SELECTED INPUT VALUES

Solar System Type
This input parameter covers different types of solar systems used for heating and cooling of buildings For example, the indicator (1) above signifies space heating with liquid collectors, collector/storage heat exchanger, fan co 's or air duct heat exchanger systems.

Fuel Type for Reference (Conventional) Heating System
Fuel types include natural gas, electricity, fuel oil, LP gas and coal. When you input an indicator (2) as above, it means electricity is the fuel used for the reference or conventional heating system.

Fuel Type for Solar Auxiliary Heating System
These fuel types are usually the same as those for the reference heating system input parameter — natural gas, electricity, fuel oil, LP gas and coal The indicator (2) represents electricity.

Collector Type
All collector types including liquid, air, evacuated tube, and others can be defined by this parameter The indicator (3) represents a liquid, flat plate, 1 cover, selective absorber collector.

OUTPUT

COLLECTOR SIZE OPTIMIZATION BY SOLCOST
Collector type = flat plate 1 glass selective
Best solar collector size for tilt angle of 55 degrees is 400 sq. ft.
Solar costs = 1000 fixed + 4800 collector + 900 storage

Input conventional system costs = 0
Initial solar investment = $6700   Down payment = $1500
Financial scenario — residence

CASH FLOW SUMMARY

Yr.	(A) Fuel/Utility Savings	(B) Maint + Insur	(C) Property Tax	(D) Annual Interest	(E) Tax Savings	(F) Loan Payment	(G) Net Cash Flow
							-1500 (Down Payment)
1	500	70	135	468	181	570	-94
2	550	73	140	459	180	570	-53
3	605	76	146	449	178	570	-8
4	665	79	152	438	177	570	42
5	732	82	158	426	175	570	98
6	805	85	164	413	173	570	159
7	886	89	171	399	171	570	228
8	974	92	178	384	168	570	303
9	1072	96	185	367	166	570	387
10	1179	100	192	349	162	570	480
11	1297	104	200	329	159	570	582
12	1427	108	208	307	155	570	696
13	1569	112	216	284	150	570	821
14	1726	117	225	258	145	570	960
15	1899	121	234	230	139	570	1113
16	2089	126	243	199	133	570	1283
17	2297	131	253	166	126	570	1470
18	2527	136	263	130	118	570	1676
19	2780	142	273	90	109	570	1904
20	3058	147	284	47	99	570	2156
Totals	28637	2086	4020	6192	3064	11400	12703

Payback time for net cash flow to equal down payment — 8 9 years
Payback time for net cash flow to equal down payment — 9 9 years
Rate of return on net cash flow — 16 3 percent
Annual portion of load provided by solar — 72 0 percent
Annual energy savings with solar system — 91 3 million btus
Tax savings = income tax rate × (C + D)
Net cash flow = A − B − C + E − F

* Similar calculations can be made for businesses and non-profit organizations where special considerations such as depreciation and tax deductions are accounted for

Fig. 6-8. An example of SOLCOST use for residential homeowner*.

EFFECT OF MOUNTING SOLAR COLLECTORS AT ANGLES OTHER THAN OPTIMUM

For space heating (only) applications, the optimum angle of a solar collector with respect to the horizon is 15° plus the local latitude. For heating-cooling applications (where collector heat is also used for cooling) latitude plus five degrees; for domestic water heating only, tilt should be equal to latitude. For Columbus, Ohio, the latitude is 40°N; therefore, the optimum tilt for heating is 40 + 15 = 55° from horizontal. For some installations it may be impractical to maintain an optimum tilt. Before deciding to install collectors at the optimum angle, it is very important that the effect on the efficiency of the collector be determined and then consider the cost of the additional collector area needed versus the expense of an elaborate frame.

Figure 6-9 graphically illustrates the effect of changing the tilt on the efficiency of the collector. The computation of the additional collector area required is also included in Figure 6-9.

EFFECT OF FACING THE COLLECTOR EAST OR WEST OF DUE SOUTH

When a collector must be oriented east or west of due south, more of the sun's energy which strikes it is lost. As in the case of changing the tilt, this requires that the collector area be increased

to compensate for this loss. The graph in Figure 6-10 will enable you to determine the efficiency of a collector which is oriented east or west of due south. For example, a collector oriented 45° east of south will be 90% efficient. Using the same formula as given in Figure 6-9, it can be seen that the collector must again be increased by 11% to offset the loss due to the orientation.

SIZING THE HEAT STORAGE UNIT

The size of the heat storage unit is dependent upon the material used to store the heat and the size of the collector. For a typical solar collector system the size of the bed storage unit is determined as shown in Figure 6-11. Section 615-7.3.1 of HUD's MPS supplement for solar heating specifies minimum storage as not less than 500 Btu per square foot of collector. Refer to Lesson 4 on converting Btu storage to gallons of water or cubic feet of rock storage.

Sizing Heat Exchangers

A third major element in many solar assisted liquid loop systems that must be sized is the heat exchanger. Recall that, in liquid loop systems using an antifreeze or special heat transfer liquid, a heat exchanger is used to separate the collector loop containing the special fluid from the storage loop which contains water. For many heating and

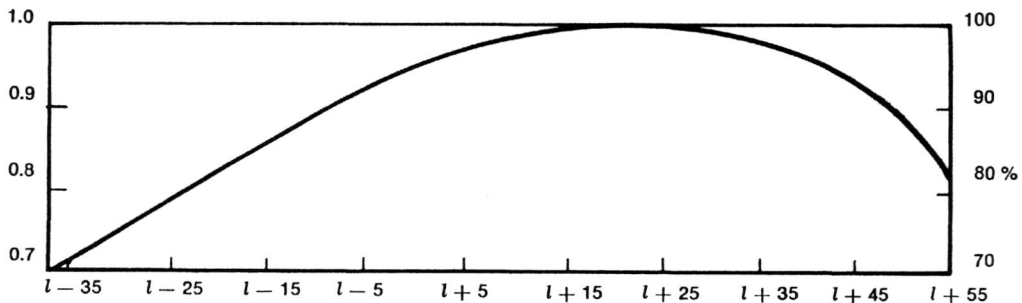

= Local Latitude

Example of How to Compute Increased
Collector Area Required

1. A collector in Columbus, Ohio, is located at a tilt of 25°.
2. Optimum for Columbus, Ohio, is 40+ 15 =55.
   Therefore, the collector tilted at an angle of $l - 15$.
3. From this angle the collector will operate at 90% efficiency.
4. To compute the increased collector area required (a) use the following formula.

$$a = \frac{1}{e} - 1$$

Where
e = efficiency of the collector (decimal)

$$a = \frac{1}{.09} - 1 = 11\%$$

Therefore, the area of the collector must be increased by 11%.

Fig. 6-9. Effect of solar collector tilt or solar heating performance.

air-conditioning technicians, this is an unfamiliar component.

Figure 6-12 shows three common types of heat exchangers. A fourth type that may be found in solar systems is an immersed element within the storage tank itself.

As noted in an earlier paragraph, the addition of a heat exchanger imposes a penalty on the sys-tem, since there must be a temperature drop through the device in order to transfer heat.

Figure 6-13 shows a simplified collector-to-heat exchanger loop. In this case, a counterflow single pass heat exchanger is illustrated. This means that the hot fluid from the collector enters the heat exchanger at the point where the heated

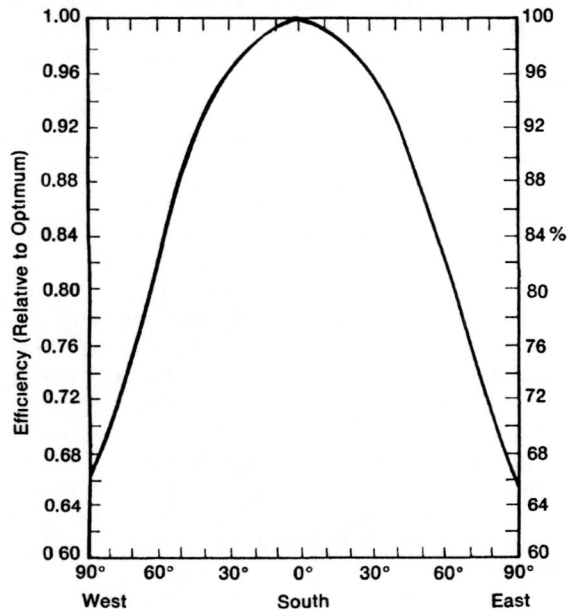

Fig. 6-10. Effect of solar collector orientation on solar heating performance.

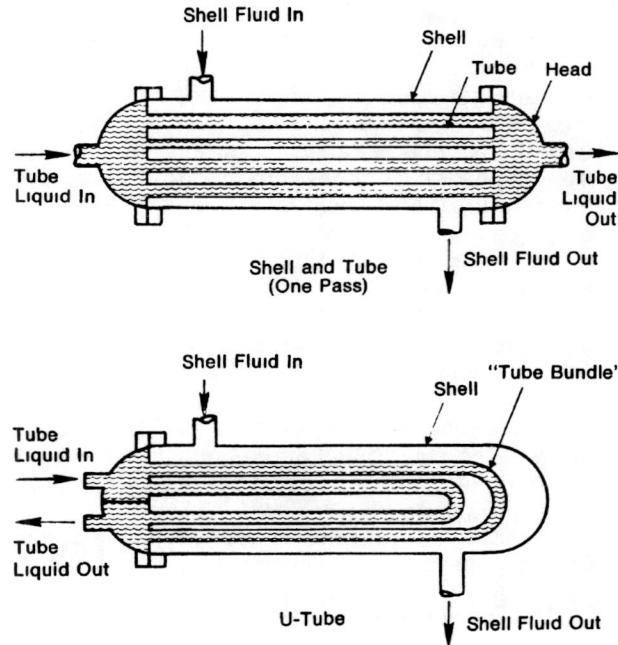

Fig. 6-12. Three basic heat exchanger types.

Air Systems
  Require ½ to 1 cubic feet of pebbles per square foot of collector area

Example. A 260 sq. ft. air collector array is installed in a house with a 50,000 Btu/h design heat loss and an overnight heat load of 287,140 in January.
  Storage required 130 (260 × 0.5) to 260 (260 × 1 0) cu ft of rock
  From lesson 4, at a specific heat of 20 Btu/cu. ft. for rock and assuming a 70 deg. F temperature rise, the storage would provide
      130 × 20 × 70 or 182,000 Btu's or
      260 × 20 × 70 or 364,000 Btu's
  Since overnight heat load is 287,140 Btu's, a 205 cu. ft pebble bed storage bin would be adequate and fall within the ½ to 1 cu. ft guideline.

Water Systems
  Require 1-2 gallons of water per square foot of collector area

Fig. 6-11. Sizing solar heating storing units relative to collector area.

water is leaving on its way to storage. This "counterflow" pattern between hot and colder fluids results in improved heat transfer.

The optimum heat transfer would occur when the temperature of the water leaving the exchanger would equal the hot fluid temperature entering the exchanger from the collector. In Figure 6-13, this suggests that the leaving water temperature of 155°F equals the collector discharge temperature. However, this is never achieved. The leaving water temperature will always be less than the hot collector fluid temperature. The difference between the entering fluid temperature from the panel and the leaving water temperature from the heat exchanger is termed the "approach temperature." Most heat exchangers in solar applications are selected based on a 10 degree approach temperature. Thus, the illustration shows a leaving heated water temperature of 145°F.

As might be expected, selections based on larger approach temperatures result in smaller, less costly heat exchangers. However, larger approach temperatures and smaller heat exchangers tend to *increase* the operating temperature of the solar panel and as described in Lesson 3, collector efficiency decreases with higher operating temperatures.

Most manufacturers of heat exchangers use computer programs to select a proper unit based on the flow and temperature conditions specified.

To select a heat exchanger, the designer would have to provide

1. the desired approach temperature (usually 10°F);

2. collector loop flow rate (from collector manufacturer specifications, but usually about .02 gpm per square foot of installed collector);

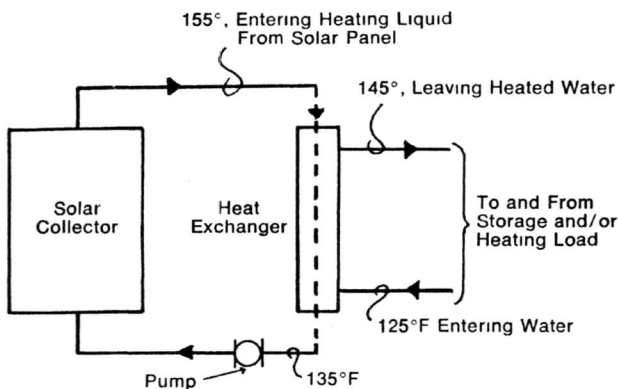

3. Btuh to be transferred (collector output at specified design radiation);

4. and either the minimum storage temperature or the flow rate from storage through the heat exchanger.

Because of the interrelationship between collector and storage and terminal unit (usually a duct coil), a complete analysis of system performance under actual conditions (laboratory or test house) must be made to arrive at one or more of the four items listed above. The designer/technician must therefore depend on a great deal of assistance from the component manufacturers to properly select the best heat exchanger.

Sizing Air Cushion Tanks

It is necessary to make provision for the increased volume of water in a liquid solar heating system caused by the heating of the fluid. There are several methods of doing this. One is to use an *open* expansion tank. This tank is open to the atmosphere and must be placed three feet or more above the highest point in the heating system.

Perhaps the more common method of providing for the expansion of the water in a closed system is the use of a closed or air cushion tank. When the system is first filled with water, a pocket of air is trapped within the tank (Figure 6-14). When the water in the system is heated, it expands and compresses the air trapped within the air cushion tank, thus providing space for the extra volume of water without creating excessive pressure.

In any system filled with water, the weight of the water develops a pressure known as *static* pressure. The static pressure at any point in the system is equal to 0.43 times the height of the column of water in feet above the point in question.

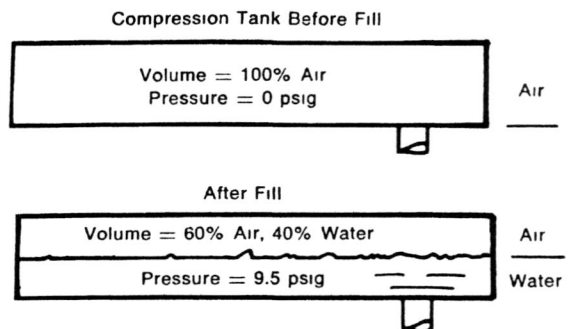

Fig. 6-13. Schematic of heat exchanger circuits.

Fig. 6-14. Cutaway view of diaphragm air cushion tank

Thus, a column of water in piping leading to or from a solar collector on a roof 19 feet above would impose a pressure of 19 × 0.43 or 8.17 psig, if a pressure gauge were connected at the base of the vertical piping.

In some solar systems, this "natural" static pressure is the only pressure in the collector loop. In others, the pressure is increased beyond this normal pressure in order to provide pressure at the *top* of the loop sufficient to prevent the hot collector discharge water from flashing into steam.

Pressure/Volume

Water, like other liquids, expands or increases in volume when it is heated. The pipes and other components of the system also expand on heating, but not enough to compensate for the increase in water volume. For a water system, the apparent expansion of the water (difference in the expansion of the water and the system components) is about 0.025 percent per degree change in temperature. In other words, if a system is filled with 50 gallons of water at 60°F, and is then heated to 200°F, the 50 gallons of water will increase in volume by 50 × 0.00025 × 140 = 1.75 gallons. Therefore, the 50 gallons of water at 60°F becomes 51.75 gallons at 200°F. The *compression tank* must permit this expansion and, at the same time, maintain the maximum pressure developed in the system below the maximum allowable pressure of the collectors and other components.

Diaphragm Air Cushion Tanks

While the conventional air cushion tank adequately protects a system from excessive pressure resulting from temperature changes within the system, its use results in air under pressure being in contact with water at relatively low temperature. This occurs at the air-water interface within the tank. Under these conditions, the water has the ability to absorb, or dissolve, some of the air in the tank. Later, as the water in the system is heated, some of this absorbed air is released from the water, since the capacity of water to absorb air decreases as the water temperature increases. This air must then either be removed by venting or it must, in some way, be returned to the air cushion tank. If it is vented from the system, the process will continue until, finally, all the air is removed from the air cushion tank and the tank becomes waterlogged. When this happens, any increase in the temperature of the water in the system results in an excessive increase in system pressure. To correct this condition, the tank must again be recharged with air.

One method of combatting the problem is the use of a *diaphragm* air cushion tank. In these tanks, an impervious diaphragm separates the water from the air, making absorption of the air by the water impossible.

Usually the diaphragm is made large enough that it can lay across the sides and bottom of the tank, as illustrated in Figure 6-15. The tanks are pre-pressurized on the air side of the diaphragm to approximately 6 psig. Under these conditions, not only is the air and water kept apart, but smaller tanks may be used.

Determining the correct size expansion tank is quite involved if basic formulas are used. Fortunately, tank manufacturers have computed simplified tables for their specific designs to ease the selection process.

Table 6-3 is one example of a simplified selection table. Here, the tank size is given in terms of gallons of tank capacity per gallon of water in the loop. Thus, a loop with a fill pressure of 12 psig for an elevation of 19 feet would require a conventional cushion tank with a capacity of 0.22 gallons of water in the loop. If a loop contains 25 gallons of water, then the expansion tank must have a capacity of 25 × 0.22 or 5.5 gallons. A diaphragm tank could be slightly smaller (.17 × 25 or 4.25 gallons).

If the fluid in the collector loop is not water, then the expansion tank may have to be increased in size if the fluid's expansion rate is greater than that of water. Table 6-4 shows correction factors for ethylene glycol (antifreeze) in various concentrations. For a 50% concentration and a 150°F design temperature, the tank size would have to be increased by a factor of 1.8 (1.8 × 5.5 or 9.9 gallons). This correction was frequently overlooked in early solar designs.

Manufacturers of specialized heat transfer fluids should provide correction factors in order to properly size the expansion tank. As in all cases, follow the manufacturer's recommendation on sizing.

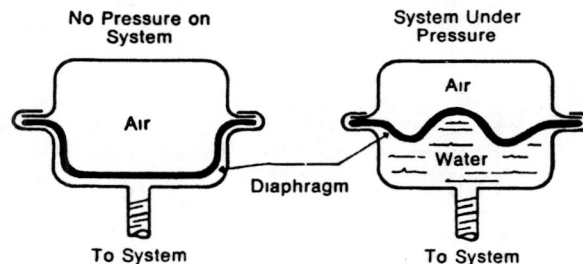

Fig. 6-15. Conventional closed expansion tank.

Sizing Air Heating Coils

Air heating coils are used to transfer heat from hot water (or steam) to air. They are used as terminal units in conventional hydronic heating systems for applications such as the heating of ventilation air. They are also used for tempering, reheating, or booster heating of circulated air for either comfort heating or process applications.

Modern air heating coils are of finned tube construction (Figure 6-16). The tubing is usually copper, and the extended surface is either aluminum or copper. The coils are enclosed in a casing designed to be installed in a duct system, so that the air being circulated is blown over the external surfaces of the coil while the hot water is circulated through the tube.

Coil ratings are based on a uniform air velocity over the face of the coil. Non-uniform air velocity may reduce the output. The output of a hot water coil used for heating air is a function of the entering and leaving water temperature, the water velocity, the entering and leaving air temperature, and the face velocity of the air entering the coil. Each manufacturer has devised its own method of presenting ratings in a set of tables, charts and/or computer selection procedures. Detailed instructions are provided for selecting coils when the key variables are given.

For comfort applications, it is possible to set some practical limits on the temperatures, flow rates, etc. needed to select a coil:

1. air velocity across the coil should range from 500 to 600 feet per minute

2. storage water (supply) temperature will range from 120 to 160°F.

3. entering air temperature will range from 68-72°F.

4. water temperature drop through the coil will range from 10 to 20°F.

5. total system air flow rate will be determined by auxiliary heating or cooling requirements.

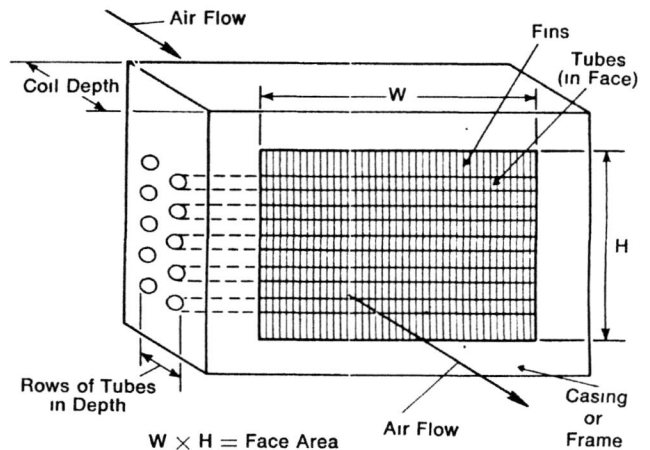

Fig. 6-16. Typical in-duct hydronic coil.

$W \times H = \text{Face Area}$

Table 6-3. Simplified expansion tank selector procedure.

Initial or Fill, Pressure psig	Max. Height of System Above Gage ft	Air Cushion Tank Capacity in Gallons Per Gallon of Water in System	Diaphragm Tank Pre Pressurized to 6 psig
4	0	0.10	
6	5	0.12	
8	9	0 15	0.11
10	14	0.19	0.14
12	19	0.22	0.17
14	23	0.26	0.20
16	28	0.32	0 24
18	32	0.39	0 29
20	37	0.48	0.36
22	42	0.63	0.47
24	46	0.85	0.64

This table is based on a final pressure of 30 psig at the low point in the system and an initial fill temperature of 60°F.

Table 6-4. Correction factors for tank sizing when anti-freeze used.

Percent of Ethylene Glycol* by Volume	Approx. Freeze Pt. °F	Maximum Design Temperature**					
		150°	160°	180°	200°	220°	240°
10%	+25°	1.15	1.13	1.1	1 09	1.08	1 05
20%	+16°	1.31	1.29	1.24	1.18	1 15	1 12
33%	0°	1.6	1.52	1.44	1 37	1.3	1.23
50%	−34°	1.8	1 73	1 6	1.5	1 42	1 36

* Interpolated from Union Carbide data book   ** Use 150°F column for temps below 150°F.

Suppose the space heating load in a solar liquid-to-air system is 70,500 Btuh. Assume further that the cooling unit requires an air flow rate of 1200 cfm.

First, calculate the required coil area.

Taking 500 fpm as the face velocity and dividing it into the total system cfm, the result is:

Total System (cfm)/Face Velocity (fpm) = Coil Area (sq. ft)
1200/500 or 2.4 square feet of coil area required.

If there is a 20°F temperature drop through the coil, the gpm of water that must be pumped from storage is

Space heating load (Btu)/[(lb/gal × min/h × specific heat(1) × Temp. Drop (°F)] = Volume of water (gpm)
70,500/(8.34 × 60 × 20 × 1) or 7.05 gpm.

Assume that the return air temperature is 70°F and the supply water temperature from storage at design conditions is 150°F. Then, referring to Table 6-5, coil A has an output of 72,800 Btu/h at these conditions. This is adequate for the purposes of this lesson.

Complete manufacturer's literature will, of course, offer many tables for coils of different areas, flow rates, and entering conditions from which to choose a unit for a given system.

SIZING FANS, PUMPS, DUCTS, AND PIPES

The sizing of fans, pumps, ducts, and pipes will be familiar to most people taking this course. For those who may need to review these procedures the following references are suggested:

Duct Design: Appendix B and C of this document

Section 3 of SMACNA Solar Standards for a detailed discussion of duct design

National Environmental Systems Contractors Association. *Equipment Selection and System Design Procedures,* 1228 17th St. N.W. Washington, D.C. 20036

Pump Selection: Appendix D of this document.

Pipe Sizing: Appendix D of this document.

NATIONAL SOLAR HEATING AND COOLING INFORMATION CENTER

Public Law 93-409 made it possible for the Department of Housing and Urban Development and the Energy Research and Development Administration to establish the National Solar Heating and Cooling Information Center. The purpose of this center is to provide a complete, one-step service facility for all information about solar heating and cooling. For example, it is possible for designers, engineers, installers, service persons, and others to obtain general information about solar heating systems and to have their name added to a categorized mailing list. Once a name is on the list, current, continuing information about various areas of interest will be mailed as it becomes available.

The center may be contacted by calling 800-523-2929 (toll free) or by writing: Solar Heating, P.O. Box 1607, Rockville, Maryland 20850.

SUMMARY

Sizing the components of a solar heating system begins with the determination of the collector size. In this unit, many techniques for determining the collector size were introduced. The simplest techniques are "rule-of-thumb" procedures. Considerably more accuracy can be achieved using data developed by manufacturers of collectors. Generally, these data are available in easy to use table form. The most complete and accurate analyses are prepared by computers based on detailed input data provided by the designer.

Life-cycle-costing was introduced as a technique for determining the most economical installation. Research data suggest that a solar heating system which will furnish 50 to 70 percent of the annual heating load will generally be the most economical installation.

The effects of mounting solar collectors at an angle other than optimum and facing them in a

Table 6-5. Sample coil output where coil area is 2.5 sq. ft. and water flow rate is 7 gpm.

Coil No Nom'l CFM	Enter Water Temp °F	Entering Air Temperature					
		70°F	84°F	88°F	90°F	94°F	98°F
Coil A 800CFM	120	34 7	25 2	22 5	21 2	18 5	15.8
	130	41 4	32 0	29 3	28 0	25 3	22 6
	140	48 2	38 8	36 1	34 7	32 0	29 3
	150	55 0	45 5	42 8	41 5	38 8	36 1
1200CFM	120	45 9	33 4	29 8	28 0	24 4	20 8
	130	54 9	42 4	38 8	37 0	33 4	29 8
	140	63 8	51 3	47 8	46 0	42 4	38 8
	150	72 8	60 3	56 7	54 9	51 3	47 7
Coil B 1600CFM	120	59 5	43 3	38 7	36 3	31 7	27 1
	130	71 1	54 9	50 3	48 0	43 4	38 8
	140	82 7	66 5	61 9	59 6	55 0	50 4
	150	94 3	78 1	73 5	71 2	66 6	62 0
2000CFM	120	60 1	43 8	39 1	36 7	32 1	27 5
	130	71 9	55 5	50 8	48 5	43 8	39 2
	140	83 6	67 2	62 6	60 2	55 5	50 8
	150	95 3	79 0	74 3	72 0	67 3	64 6

direction other than due south were presented. Graphs were provided which enable the designer to determine the reduction in efficiency of the collector for both of these conditions.

Sizing procedures for the heat storage units, heat exchangers, air cushion tanks, heating coils, fans, pumps, ducts, and pipes were discussed. Several references were cited to provide additional information about sizing each of these components.

OPERATION OF SOLAR HEATING SYSTEMS

Collecting, storing and using solar energy for space heating requires the control of air or liquid flow (or both). In Lesson Five, the various types of controls and controlled devices that are likely to be found in most solar heating systems were presented.

In this lesson, the basic operating modes of solar heating systems will be covered. Those devices and procedures reviewed should be considered the most common approaches. Quite obviously, there exists any number of highly specialized proprietary control configurations and more are sure to follow. Therefore, do *not* assume that what is learned here is the only way solar system operating cycles may be accomplished.

ALL-AIR SYSTEMS

There are four essential operating modes for a solar assisted air system: (1) space heating directly from collector, (2) space heating from storage, (3) space heating from auxiliary, and (4) storing heat. A fifth mode would be domestic water heating, if so equipped.

As noted in Lesson Five, a two-stage room thermostat is typically (but not always) used to sense space temperature and initiate demand. The first stage usually operates the solar system and the second stage operates the backup heating system.

Heating Directly From the Solar Collectors

When the sensor in the discharge air stream of the solar collector indicates that the collector temperature exceeds a pre-determined point, and the room thermostat is demanding heat, then the operating mode is as shown in Figure 7-1.

The duct arrangement shown in Figure 7-1 indicates a two-blower system. This is fairly typical, although a single blower configuration is also used (Figure 7-2). More exotic designs may even include three blowers. Also, for simplicy, manual and back draft dampers are not illustrated.

For the arrangement illustrated, dampers J-1, J-2 would be positioned to isolate the storage circuit, damper J-3 would be opened, and the economizer would be closed. The air flow path would follow the shaded area, provided collector flow rate and auxiliary heating air flow requirements were the same. If there is a difference in air flow requirements, then the bypass damper J-4 would be opened and some portion of the air flow would circulate through the bypass duct. Thus, if collector air flow was 800 cfm and the auxiliary furnace/AC unit required 1200 cfm, 400 cfm would flow through the bypass duct. In this mode, both the furnace fan and collector fan are operating.

Heating From Storage

When the collector air temperature is below a

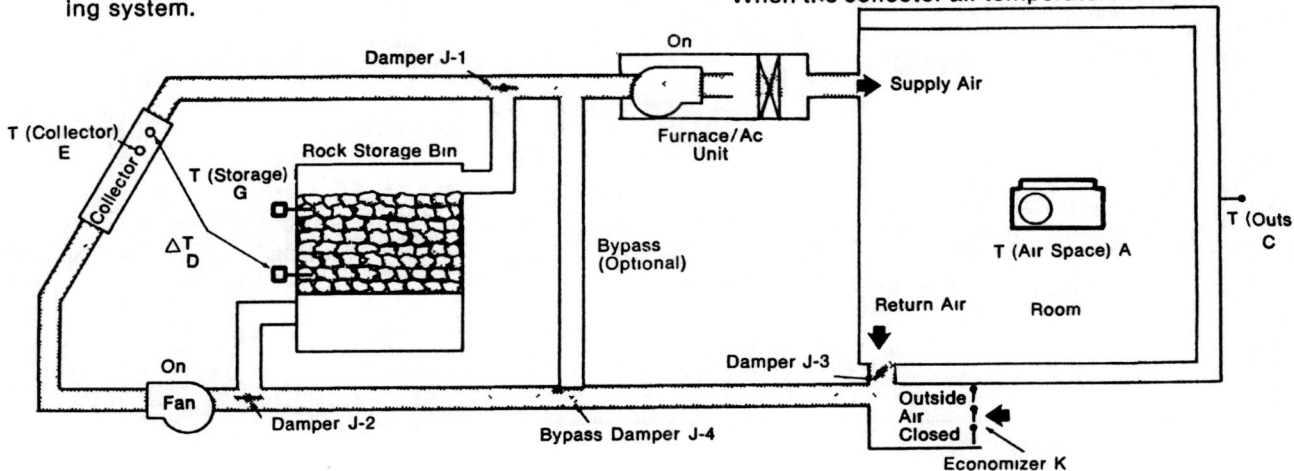

CONDITIONS:
1. Airspace requires heat (A).
2. Collector temperature (E) exceeds control point

Fig. 7-1. Heating directly from collectors of an All-Air System.

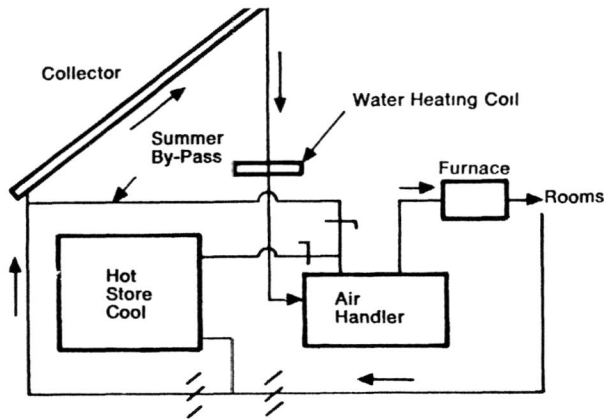

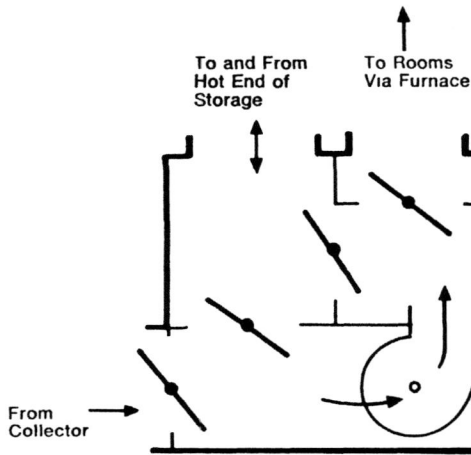

Fig. 7-2. Air handler, single blower system design. (Sketch above shows use in complete system).

predetermined set point, and the storage temperature is above a specific point (e.g., 90°F or above), then dampers J-1 and J-2 are repositioned to direct the air through the rock storage unit. At this time, the collector fan would shut off. (See Figure 7-3.)

Heating with Auxiliary Furnace/AC Unit

In most instances, when the temperature drops in the rock storage and fails to satisfy space heating needs, the space temperature will drop and the second stage of the room thermostat will close to start the auxiliary heating. This unit may be a gas, oil or electric furnace or perhaps a heat pump. Figure 7-4 illustrates the flow path for this mode.

Storing Heat

When there is no demand for heating from the first stage of the room thermostat but there is collectable heat, the system switches to the heat storage mode. Figure 7-5 shows the flow path. In this mode, dampers J-1 and J-2 are positioned to route the discharge air from the collector through the pebble-bed storage. The collector fan is on and the auxiliary furnace fan may be either on or off, depending on the use of CAC (Continuous Air Circulation) or cycle fan operation.

Please note that the flow path for storing heat is reverse of the path used to remove heat during the heating from storage operating mode.

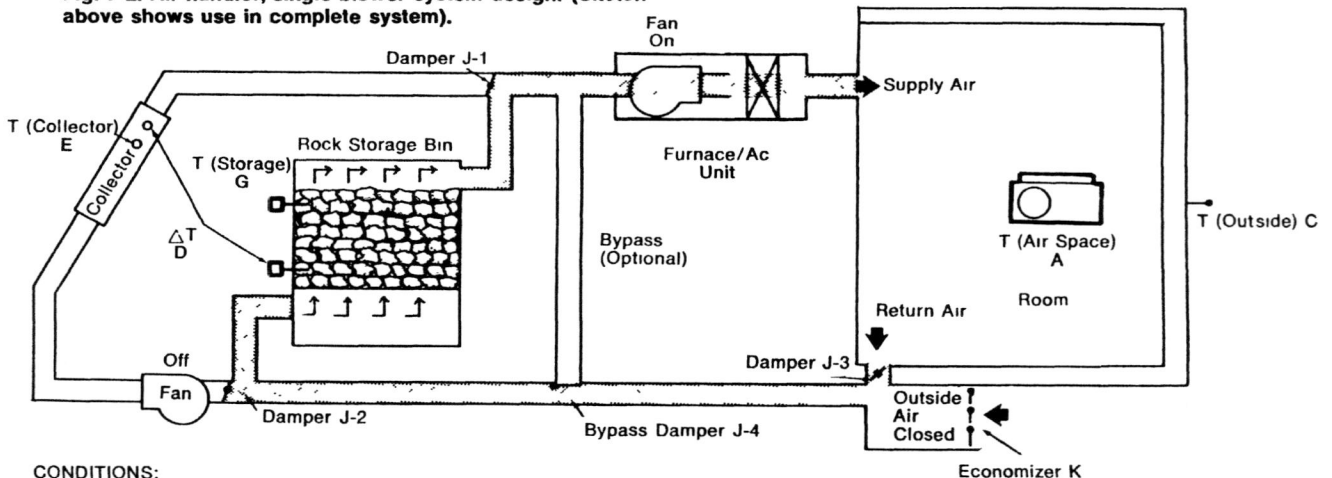

CONDITIONS:
  1. Airspace requires heat (A).
  2. Collector temperature (E) is below control point.
  3. Storage temperature (G) is above control point.

Fig. 7-3. Space heating from storage.

7-2

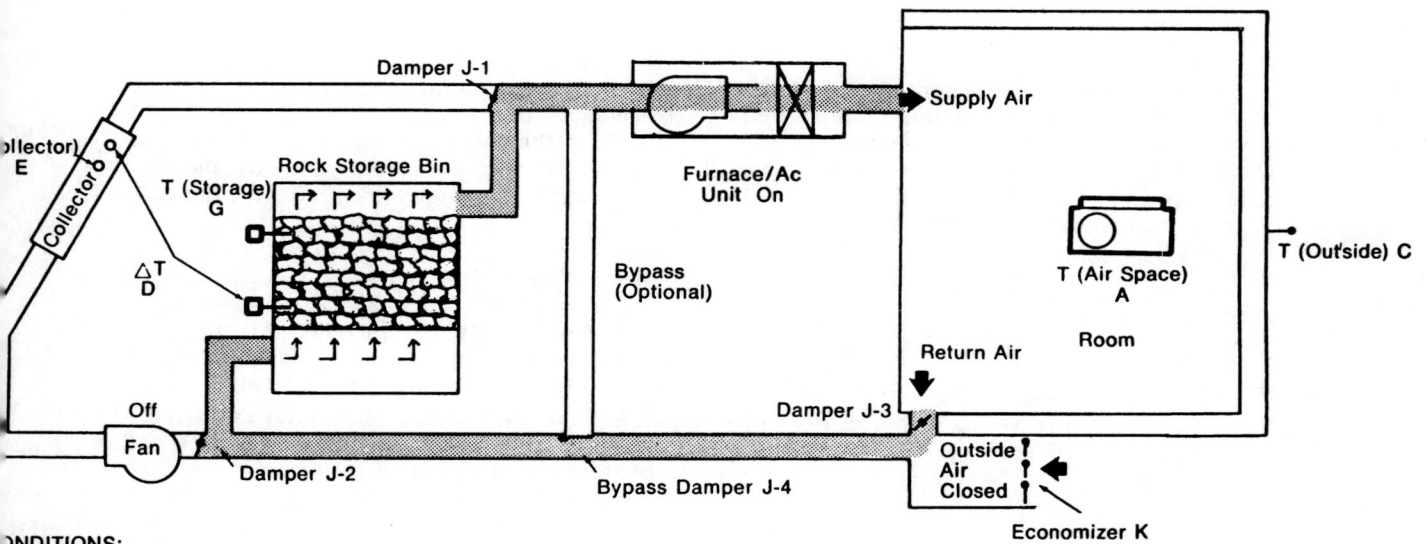

CONDITIONS:
1. Airspace requires heat (A).
2. Collector temperature (E) is below control point.
3. Storage temperature (G) is below control point.
4. Furnace is energized by 2nd thermostat stage (A).

Fig. 7-4. Heating using the auxiliary heating unit.

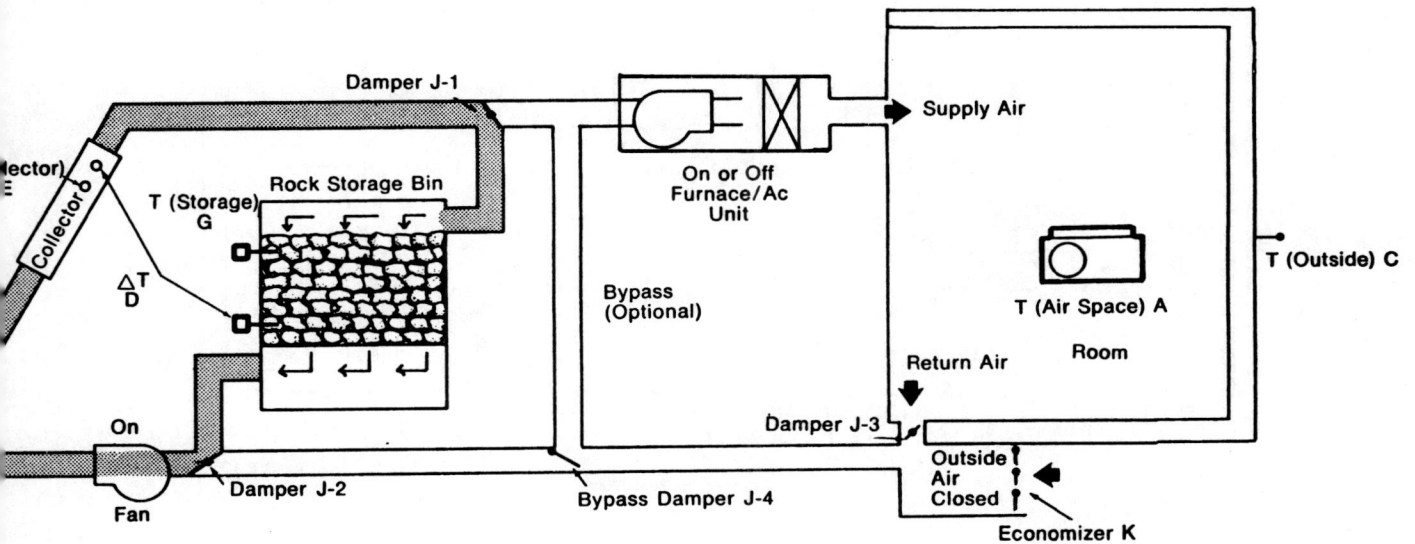

CONDITIONS:
1. Airspace doesn't require heat (A).
2. Collector temperature exceeds storage temperature (D).

Fig. 7-5. Storing heat.

This flow path direction is essential to take full advantage of the temperature stratification which occurs in the rock storage. This permits the hottest air to be utilized when heating from storage.

In a typical operation, the collector fan would be turned on whenever the temperature difference between the air entering and leaving the collector exceeded 20 degrees, regardless of a demand for heat from the room thermostat. Dampers automatically adjust for the storage mode and heat storage continues until the room thermostat calls for heat.

Domestic Water Heating

Preheating of domestic water is possible in an all-air solar system by the addition of a fin-tube heat exchanger in the air stream leaving the collector. Figure 7-6 illustrates the concept.

A small circulator pumps water through the coil and into a water storage tank. Preheated water from the storage tank is then drawn into the conventional DHW heater. The pump is turned on whenever the collector fan is on and the temperature in the water storage tank is less than about 140°F.

Summer Operation. Figure 7-7 illustrates a possible operating mode where conventional cooling coupled to an economy cycle might be employed.

If the room thermostat calls for cooling, the bypass damper J-4 is positioned to direct return air through the bypass duct into the furnace/air conditioning unit. The collector loop and storage circuit are now isolated. Conventional cooling can then be accomplished.

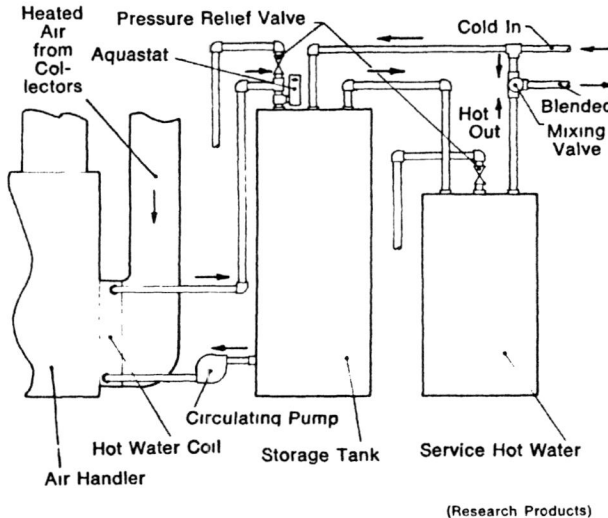

(Research Products)

Fig. 7-6. Preheating domestic hot water with an air system.

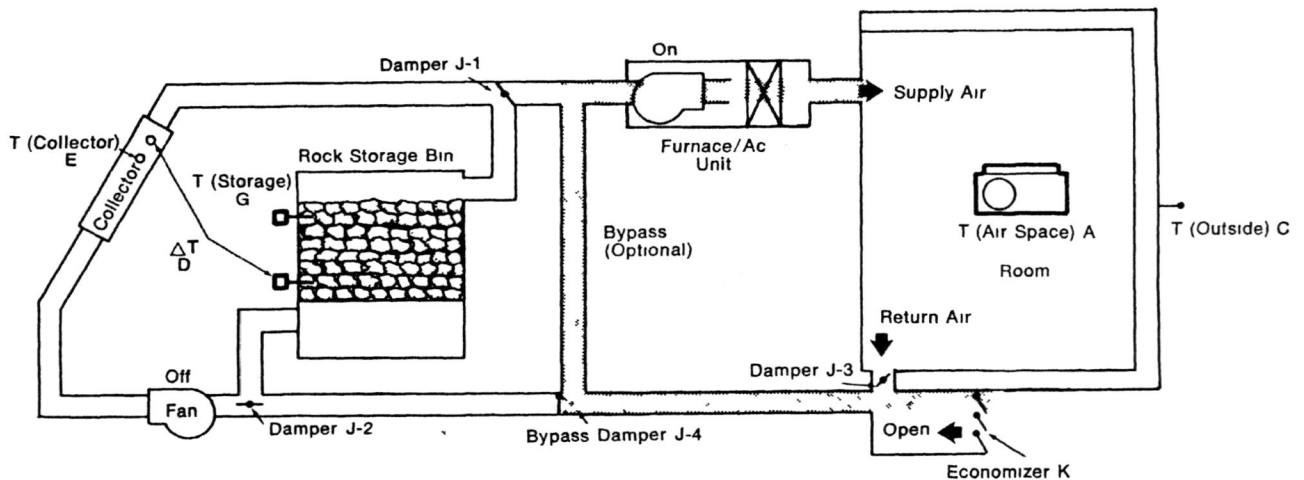

CONDITIONS:
1. Airspace requires cooling (A).
2. Outside conditions (C) are satisfactory for providing cooling with economizer (K).
3. If load cannot be met with economizer (K), conventional cooling system will be energized from (A).

Fig. 7-7. Summer cooling mode.

Whenever outdoor temperature and humidity are low enough, the economizer dampers will open. Then, outside air will be used to provide space cooling rather than the mechanical refrigeration unit.

Also, during summer months, it is possible to use the collector loop to provide preheating of domestic water if a second bypass duct is installed so that the storage loop is isolated. Figure 7-8 illustrates this arrangement.

In this mode, a manual damper, J-6, is open in the summer and a manual damper, J-5, in the storage loop is closed. At the start of the winter heating season, these dampers must be changed; J-6 is closed and J-5 is opened. Another design (not shown) uses outside air to supply collectors and, after passing through the water coil, the air is discharged to the outside. Figure 7-9 shows one type of air handler used in an all-air system including the DHW coil.

To prevent scalding water from entering the domestic hot water line, it is necessary to install a thermostatic mixing valve in the pipe from the hot water heater. The themostatic mixing valve is connected to the cold water supply piping and the hot water line (refer to Figure 7-6). The valve's purpose is to mix cold water with overheated water to insure that the hot water temperature will remain nearly constant at the faucet. The thermostatic mixing valve will be particularly important during the summer months when solar energy is plentiful.

In case the water were to overheat, a temperature/pressure (T/P) safety relief valve is installed at the top of the solar heated water storage tank. A drain line from the T/P valve to a point near a floor drain will prevent escaping water from causing any damage.

IMPORTANT DESIGN AND INSTALLATION FACTORS

Blower power and air leakage are the most important factors in the design and installation of an air solar heating system. A well-designed system will have an equal pressure loss in the collector and the storage of approximately 0.3 inch of water. The addition of ducts and filters could increase the total pressure loss for the system to as much as one inch of water. Because this pressure loss is about double that found in a conventional forced air heating system, larger blowers are required. The process of storing energy requires that the blowers run more hours per day. Therefore, a one inch of water pressure loss is the maximum acceptable from the standpoint of blower operations cost.

Loss of heat and pressure due to air leakage is more of a problem in an air solar heating system than a conventional forced air heating system.

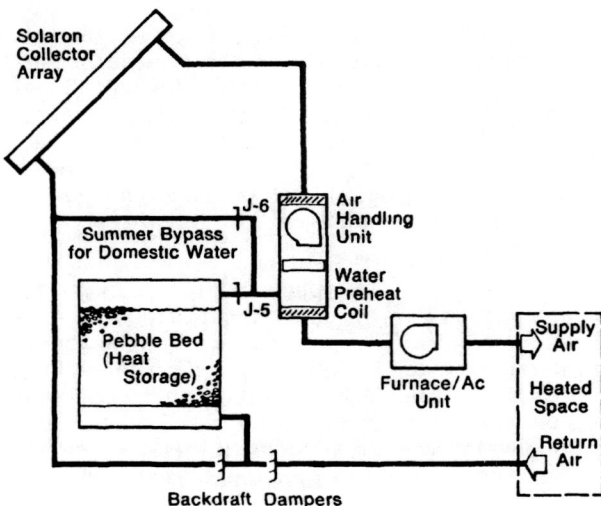

Fig. 7-8. Summer domestic water heating with an all-air system.

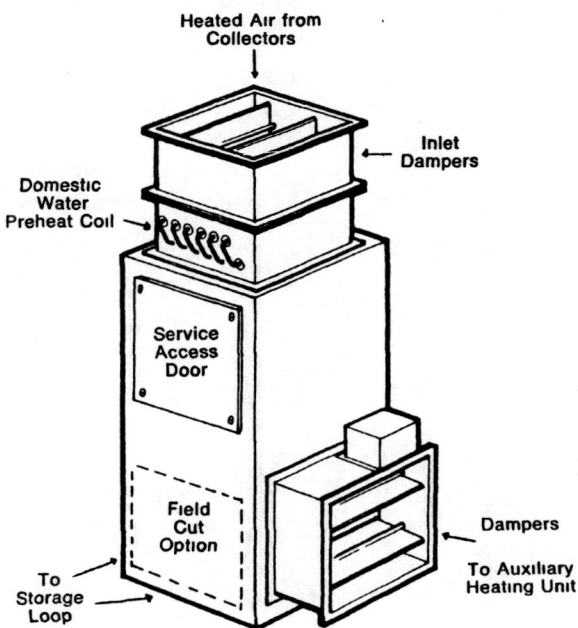

Fig. 7-9. Typical all-air solar air handler.

This is due to four different factors: (1) air pressure in the system is higher, (2) there is more ducting, (3) the system runs more hours, and (4) there may be more ducting through unheated space. To prevent leakage, ducts must be taped well at all joints. Fiberglass board or fiberglass lined sheet metal ducts should be installed which have an insulation valve of R-4. This will reduce the heat loss through the ducts to an acceptable level.

Duct sizing should be based on an air velocity of 700 to 1000 feet per minute (fpm). All duct bends should be fitted with turning vanes to reduce pressure losses.

Blowers should be forward-curved squirrel cage type. To achieve quiet operation, blowers should be belt driven at 900 to 1700 rpm. Flexible connections between blowers and ducts will add greatly to the quietness of operation. Direct driven blowers with the motor in the air stream have a shorter service life than externally mounted motors.

Dampers fitted with live rubber seals are recommended for positive shut off and smooth operation. All damper drive motors should be located on the outside of ducts with direct coupling to the damper shaft through flexible linkage. Additional safety and reduced wiring cost can be achieved by installing low voltage (24) damper motors which have spring returns.

Back draft dampers may be of the shutter or flat type. They must be mounted to provide a positive seal against reverse air flow.

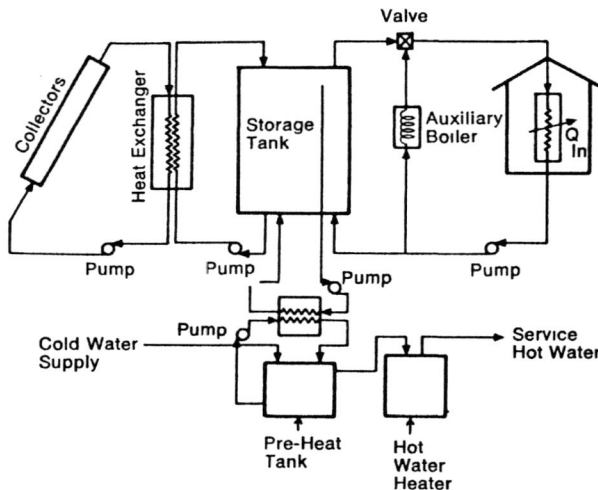

Fig. 7-10. A representative solar space and service hot water system without direct heating from collectors.

HEATING WITH A HYDRONIC SOLAR HEATING SYSTEM

Liquid (hydronic) solar heating systems operate basically the same way as air solar systems. The most basic difference is the addition of a liquid-to-water heat exchanger in closed systems. Also, some liquid systems are arranged so that the modes of operation are simply (1) storing heat, (2) heating from storage, and (3) heating by auxiliary. Space heating directly from collectors is not possible. (Figure 7-10 illustrates one example.)

One disadvantage of systems that eliminate direct collector heating is that collectors must always operate above storage tank temperature and, hence, at a higher inlet temperature. This lowers collector efficiency.

In systems where heating from the collector is provided, lower temperature water can be used to effectively heat the space, thus increasing energy utilization. To take full advantage, however, the terminal (space heating) device must also be able to utilize low temperature water to heat successfully. This usually rules out hydronic baseboard units sized for 220°F water, and many liquid systems use duct coils to transfer heat to circulating air. (Radiant floor panels operating at 110°F are another possible option.)

Figures 7-11 through 7-13 illustrate a typical hydronic-to-air solar heating system with most specialty items omitted for clarity. Accompanying Figure 7-11 is a table which summarizes the condition of each pump and blower in the system for each operating mode.

Sensors in the collector and solar heated water storage tank, along with a thermostat located in the space to be heated, provide the necessary information to a central control unit. Based on the information (electrical signals) received, the central control unit opens or closes valves and activatees the appropriate pumps and/or blowers. This control system makes the system completely automatic.

Heating from the Collector

In Figure 7-11, space heating is being accomplished from the collectors. The room thermostat has indicated a need for heat and the sensor in the collector is above some predetermined point, for example 90°F or higher. The collector pump and storage pump are turned on and the liquid/water flow is as indicated through the duct coil. The furnace or air handler blower is also on to circulate room air over the now hot coil.

Heating from Storage

In Figure 7-12, the room still requires heating

but the collector temperature is below the control point while the storage temperature exceeds its control point, again perhaps 90°F. Both storage and collector pumps are stopped and the load pump is started to take heat from storage and circulate it through the duct coil.

Heat Pumps Used for Auxiliary

As noted in Figures 7-11 and 7-12, if a heat pump is used for auxiliary heat, the duct coil should be placed ahead of the heat pump refrigerant coil. This is necessary to avoid higher head pressure on the compressor due to higher temperature air moving through the refrigerant coil. But with the duct coil in the supply system, a higher cut-off temperature, perhaps 120°F vs.

90°F, must be used for the return side coil. Raising the minimum "draw down" storage temperature reduces the overall efficiency of the solar system.

Storing Heat

Figure 7-13 illustrates the heat storing mode. Space conditions are satisfied, but the temperature differential between the collector and storage is high enough, 20°F for example, to initiate collector and storage pump operation. Diverting valve one is activated to divert the discharge from the heat exchanger into the top of the storage tank. The storage pump removes cooler water at the bottom of the tank and recirculates it through the heat exchanger to be heated.

Operating Mode	Storage/Collector Pumps	Load Pump	Auxiliary Heater	Blower
Storing Heat	On	Off	Off	Off
Heating from Storage	Off	On	Off	On
Heating by Auxiliary	Off	On or Off	On	On

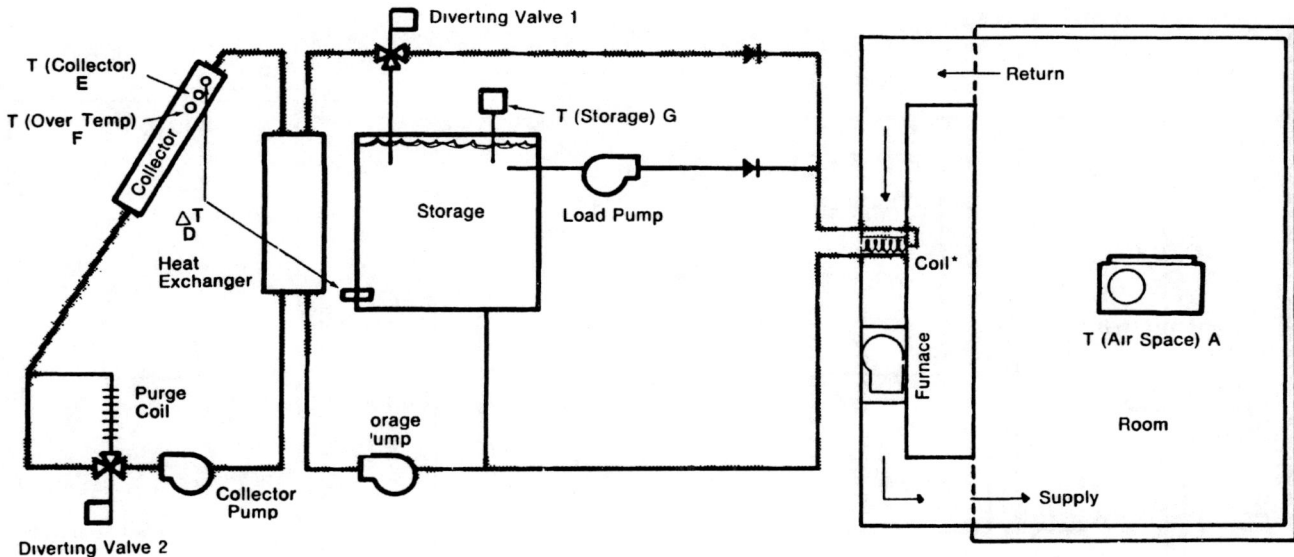

* NOTE If heat pump is used, solar coil is placed downstream from heat pump coil.

Conditions·
1. Airspace requires heat (A).
2. Collector temperature (E) exceeds control point.

Fig. 7-11. Heating from the collector.

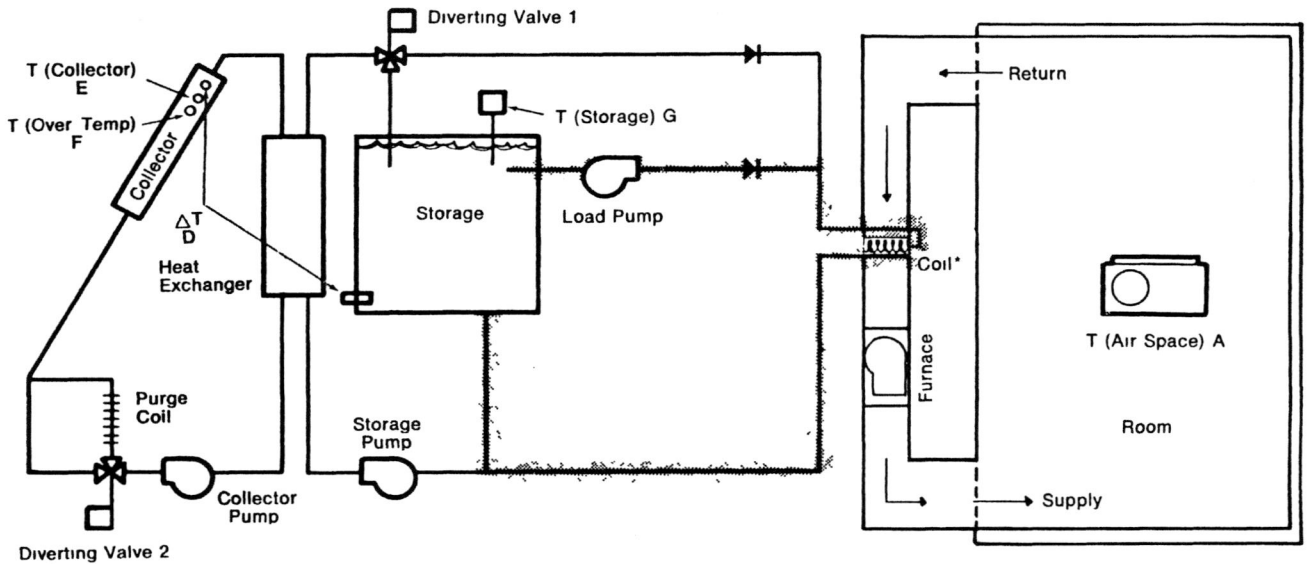

* NOTE: If heat pump is used, solar coil is placed downstream from heat pump coil.

CONDITIONS:
1. Airspace requires heat (A).
2. Collector temperature (E) is below control point.
3. Storage temperature (G) exceeds control point.

Fig. 7-12. Heating from storage.

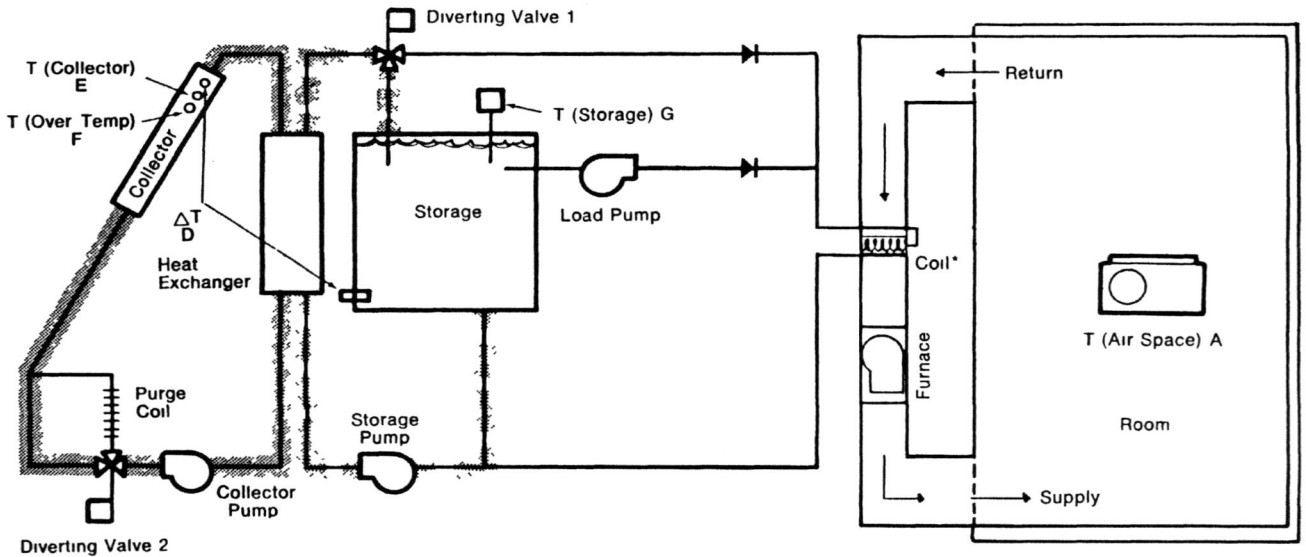

* NOTE: If heat pump is used, solar coil is placed downstream from heat pump coil.

CONDITIONS:
1. Airspace does not require heat (A).
2. Collector temperature exceeds storage temperature $\triangle$T (D)

Fig. 7-13. Heat storage mode.

CONTROLLING THE SEQUENCE OF OPERATION

These operating modes will now be described in greater detail using a control schematic along with a system diagram as shown in Figure 7-14.

First Stage Heating from Collector

Thermostat TSI positions valve VS through R1 to direct flow to the load coil. Collector pump P1 and storage pump P2 are enabled to run through relay R1.

If the collector plate temperature TP is greater than 90°F (adjustable), then pump P1 will operate through relay R3 (in series with R1). A time delay will delay pump P1 shutdown to prevent short-cycling. Pump P2 operates only when pump P1 is operating and is controlled by relays R1 and R3.

Valve VS is also under the control of the high limit air temperature sensor TA through relay RA to divert flow from the coil if the discharge air temperature exceeds the set point (make 140°F, break 120°F). If first stage heating is satisfied, valve VS is positioned to divert flow to the storage tank. Relay RF through sensor TF will inhibit the

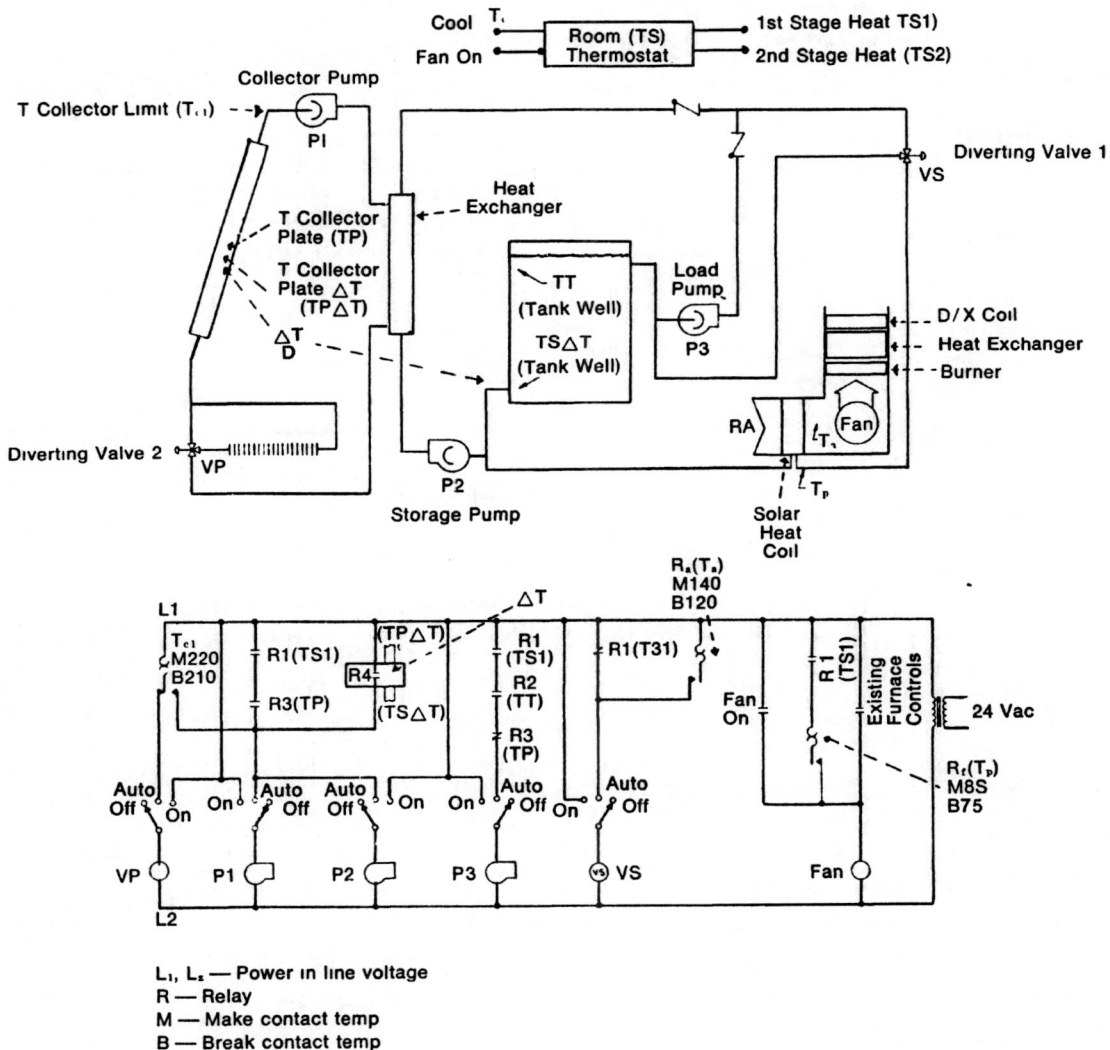

L₁, L₂ — Power in line voltage
R — Relay
M — Make contact temp
B — Break contact temp

Figure 7-14. System control schematic.

7-9

fan until the fluid temperature is at least 85°F (make 85°F, break 75°F). The furnace fan operates in first or second stage.

First Stage Heating from Storage

Heating from storage is accomplished whenever energy is not available from the collectors, a heating demand occurs and energy is available from storage.

On a call for heat, thermostat TSI through relay R1 will enable pump P3 to operate. Pump P2 is interlocked with pump P3 such that P3 will operate only when P2 is not operating. Pump P3 will operate if the storage tank temperature TT is greater than 90°F (adjustable) through relay R2 which is in series with R1.

Second Stage Heating

If first stage heating cannot be satisfied, then thermostat TS calls for second stage heating. First stage heating continues during second stage heating.

Cooling

Conventional controls and sequence are used for air cooling.

Storage Tank Charging

Charging is accomplished by diverting flow to storage through valve VS. Storage can occur only when there is no call for heating. When there is no call for heating, pump P1 and hence pump P2 are under the control of the differential temperature sensor Delta T through Relay R4. Pump P1 can run when TP is greater than TS by ten degrees Fahrenheit (adjustable). The time delay prevents short-cycling of pump P1.

In some installations, a purge cycle is used to prevent overheating of the collectors. A number of manufacturers include a purge or dump cycle to discharge unwanted heat into the atmosphere. In Figure 7-15, one possible purge cycle is illustrated. Both storage and load pumps are off and the collector pump is operating. Diverting valve 2 is activated and diverts heated water through the purge coil to be dissipated into the atmosphere.

Domestic Water Heating

Preheating of domestic hot water involves yet another operating mode. One example is shown in Figure 7-16. This is actually a sub-circuit of the system shown previously in Figure 7-10.

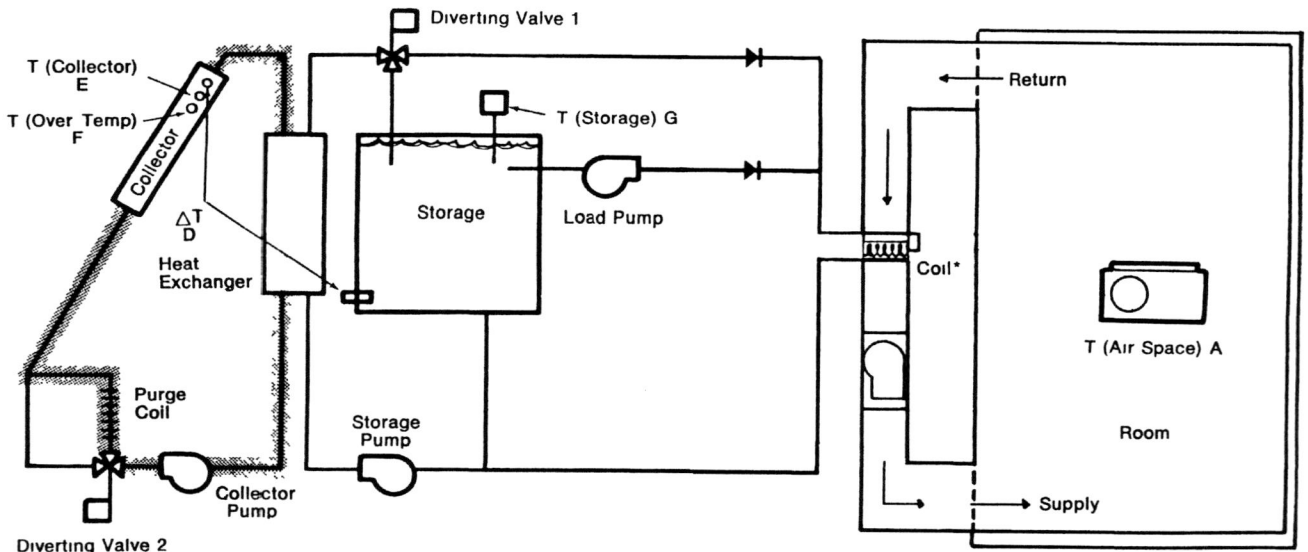

* NOTE: If heat pump is used, solar coil is placed downstream from heat pump coil.

CONDITION:
1. Over temperature condition exists in system (F).

Fig. 7-15. Purge cycle discharges unwanted heat to the atmosphere.

The heat exchanger pump and the preheat tank pump are operated whenever there is a difference in temperature between the storage tank and the preheat tank. When the preheat tank reaches a prescribed maximum temperature, the pumps are shut down. A conventional gas, electric, or oil fired water heater is used to provide additional heat whenever required.

Solar Assisted Heat Pump

Heat pumps have been widely accepted in recent years as a means of reducing total energy cost. Conceptually, heat pumps use the refrigeration cycle to function, thus a single installation can provide both heating in the winter and cooling in the summer. Figure 7-17 illustrates how the heat pump functions to provide heat.

Figure 7-17 illustrates how the heat pump operates in the cooling mode. Put simply, the basic difference is that, in the first case, the heat exchanger inside the building is used as a condenser and gives off heat. In the second case, the heat exchanger inside the building is used as an evaporater and absorbs heat.

The installation of a heat pump with a solar heating system introduces several interesting possibilities. The simplest installation would be to use a conventional automatic heat pump as the auxiliary heat unit in a hydronic system. This is shown in Figure 7-19.

One current arrangement includes these operating modes:

1. In mild weather, solar heated duct coil provides heating.

2. As temperature drops, for example to 45-50°F, the heat pump supplies heat and the solar unit goes into heat storing mode.

3. At heat pump balance point (output of heat pump equals building load), duct coil is supplied heat from storage to supplement heat pump output.

4. In the event of no solar heat at collectors or in storage, electric resistance heaters are energized to supplement the output of the heat pump.

Another approach is to use a conventional water-to-air heat pump with solar assist. This is shown in Figure 7-20. There are three operating

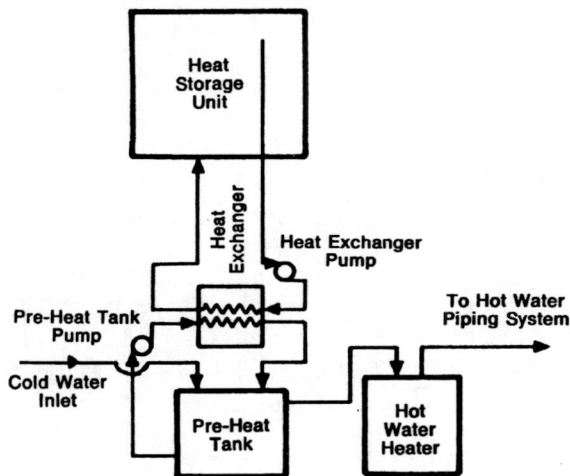

Fig. 7-17. Heat pump in heating mode.

DOMESTIC HOT WATER PREHEATING SUB-SYSTEM

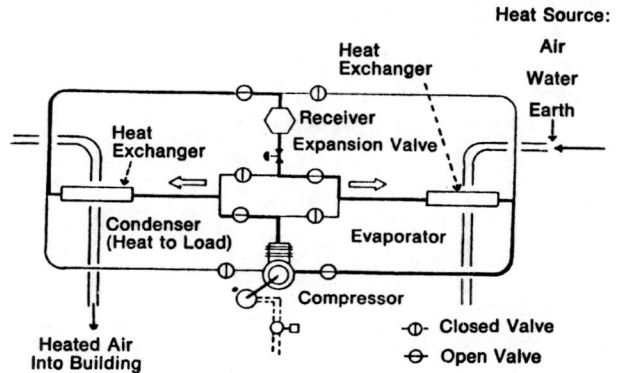

Fig. 7-16. Domestic water preheating sub-system.

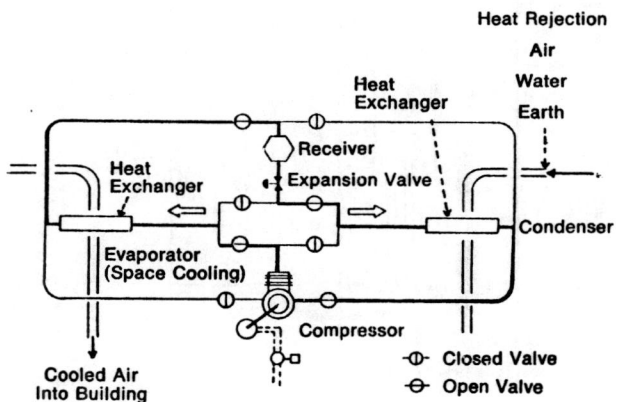

Fig. 7-18. Heat pump in cooling mode.

modes for this system:

1. Storage water 80°F or above, solar heating through duct coil.

2. Storage water between 80°F and 45°F, heat pump extracts heat from storage, boosts temperature via refrigeration cycle and supplies indoor refrigerant coil placed in duct.

3. If storage temperature falls below 45°F, auxiliary heat raises the temperature of the storage tank to 55°F.

In this case, the use of the water-to-air heat pump permits greater drawdown of storage temperature which, as noted earlier, improves overall collector performance. Whether the improvement is great enough to offset additional operating time and electricity is not yet well established.

A third combination of heat pump and solar heating systems is shown in Figure 7-21. In this instance, a liquid-to-air solar system supports an air-to-air heat pump.

The solar system serves two duct coils. One is in the indoor air-handler for direct space heating; the other heats air supplied to the "outdoor" section of the heat pump. Descriptions of the three operating modes follow.

1. **Heat Pump Heating with Solar Preheat Assist**

a. room thermostat demands heat

b. storage temperature too low for direct heat supply to indoor duct coil

c. outside temperature is lower than storage temperature

d. pump from storage and diverting valve supply heat to preheat coil for heat pump outdoor section

As in the water-to-air temperature example of Figure 7-19, this preheating of outdoor air, which is the heat source for the heat pump, raises the coefficient of performance of the heat pump. This permits a greater drawdown of storage temperature which improves collector performance. Since cold outdoor air moves over this coil, freeze protection may be required.

2. **Heat Space Directly From Storage**
Whenever storage temperature exceeds control point, for example 90°F, heat is pumped directly from storage to duct coil for direct space heating.

3. **Heating Space With Heat Pump Only**
When outside air temperature is higher than

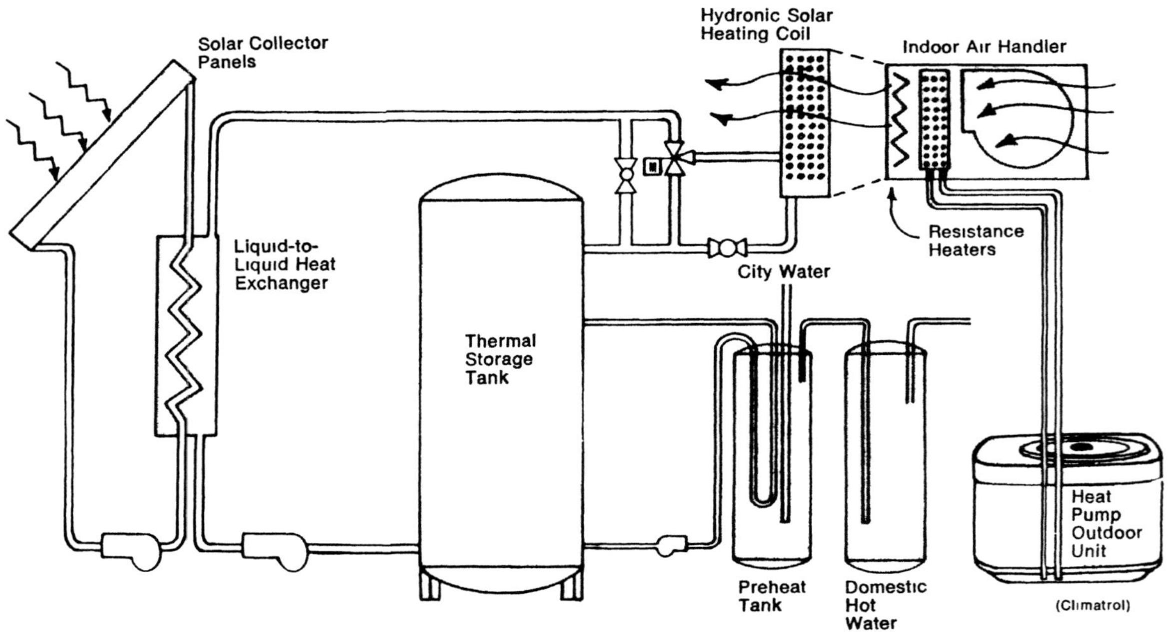

Fig. 7-19. Air to air heat pump with solar assist.

storage temperature, the heat pump operates alone, extracting heat directly from outside air. In any of these modes, if the collector temperature exceeds storage temperature, then the differential controller will start the collector pump and begin charging storage.

To conclude this review of common operating modes, the charging cycle (heat storage) will be examined in closer detail.

Figure 7-22 shows a graph of collector and storage tank temperature (vertical scale) versus time of day. At the start of the day, the pump is *off* and the tank temperature is higher than the collector temperature. But as the sun rises, the collector temperature rises sharply and is soon higher than the tank temperature. Typically, when the collector reaches point 1 in the figure, or perhaps 20 degrees higher than the tank temperature, the differential thermostat starts the collector pump.

But, as the pump starts, there is a drop in collector temperature caused by the inrush of cool liquid; this is point 2. Now, if the on and off set points for the differential thermostat are set too close (e.g., 10°F on, 5°F off), there is a chance that the pump will short-cycle. That is, it will start when the collector temperature is at point 1 and stop immediately when the collector temperature temporarily drops to 2.

During the day, collector and storage temperatures rise until late afternoon. By 6 p.m. or so, the collector temperature reaches point 3, and the collector-to-tank differential is at the *off* set point of 3°F, for example.

When the pump is stopped, the temperature of the collector will increase since no heat is being carried away and some solar energy is still striking the collector. This rise is shown as point 4. Again, if the rise in temperature is sufficiently high and the *on* temperature differential suffi-

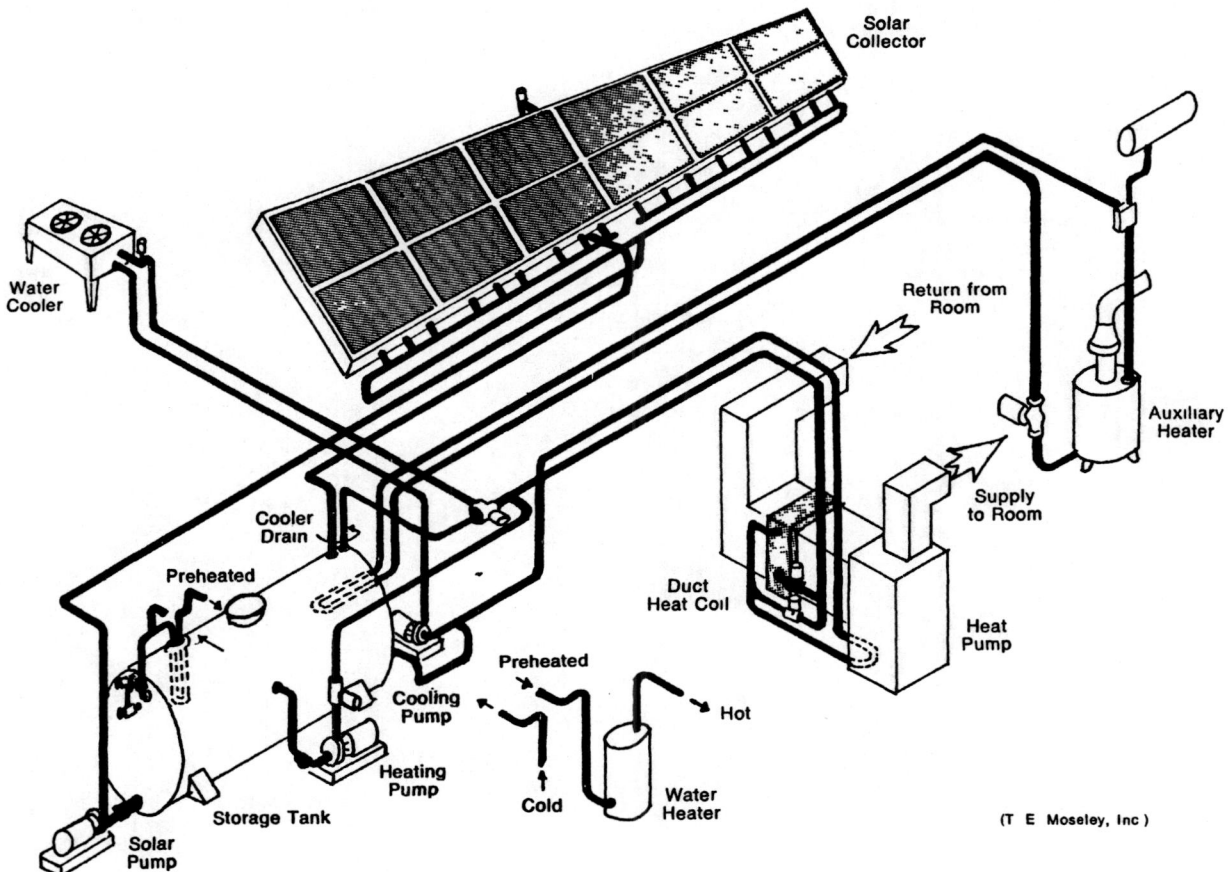

Fig. 7-20. Water to air heat pump with solar assist.

7-13

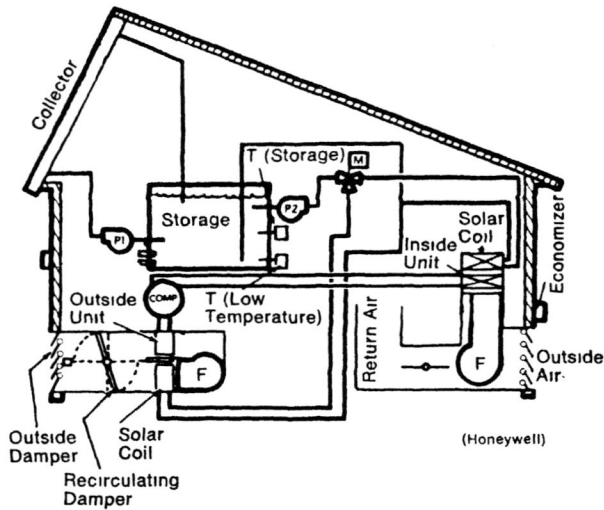

Fig. 7-21. Liquid to air solar system outfitted with an air to air heat pump.

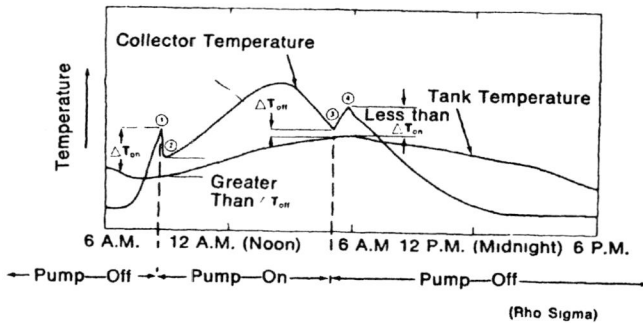

Fig. 7-22. Pump operating cycle in storage "charging" mode.

ciently small, then the pump would start up again and begin short cycling. To avoid this, the *on* differential must of course be greater than the collector-to-tank differential at point 4.

SUMMARY

In this lesson, the more common types of solar air and liquid systems have been presented. The relationship of each of the basic components and the controlling mechanisms were discussed in some detail. The basic modes of operation discussed are:

1. Heating from the collector.

2. Heating from storage.

3. Heating an auxiliary furnace.

4. Storing heat.

5. Preheating domestic hot water.

In addition, the possibilities of using a heat pump in combination with a solar system were discussed.

DOMESTIC WATER HEATING

Solar heating of domestic water can be accomplished by including the appropriate equipment like those discussed in the two previous units in solar heating systems. Also, it is possible to install a solar water heating system, Figure 8-1, without a complete heating system. In fact, solar water heating is the oldest domestic use of solar energy.

Heating water with solar energy generally makes more "sense" economically than whole house space heating, because hot water is required all year long. The opportunity to obtain a return on the initial investment in the system each and every day of the year is a distinct economic advantage. Only moderate collector temperatures are required to cause the system to function effectively. Thus, the heating of domestic water can be accomplished during less than ideal weather conditions.

This lesson is organized to provide: (1) an overview of the different types of solar water heating systems, (2) an explanation of how each of these systems function, and (3) discuss location and sizing of the collector for a basic system. Some of this discussion will be similar to the information presented about solar heating systems in Lesson Seven because a domestic water heating system is, in effect, a small solar heating system.

Types of Solar Domestic Water Heating Systems

There are basically three different types of solar domestic water heating systems: (1) direct heating/thermosiphon, (2) direct heating/pump circulating, and (3) indirect heating/pump circulating. The following three sections of this lesson treat each of these systems independently.

Direct Heating/Thermosiphon Circulating System

This is the simplest form of the solar domestic water heating systems because it requires no pumps and very few valves. The flow of water through the collector and storage tank is controlled by the temperature of the water in the collector. The warmer the water becomes, the more rapidly it rises through the collector and enters the storage tank which is elevated above the collectors. Cool water is heavier (has greater density) and flows from the bottom of the storage

(Grumman)

To heat domestic water requires considerably less collector area than for space heating.

Table 8-1. Daily hot water usage (140°F) for solar system design*

Category	One and Two Family Units 1/ and Apts. up to 20 Units					Apts. of 2/ 20-200 Units	Apts. of 2/ Over 200 Units
No. of People	2	3	4	5	6	—	—
No. of Bedrooms	1	2	3	4	5	—	—
Hot Water/Unit (gal./day)	40	55	70	85	100	40	35

1/ Assumes 20 gal. per person for first 2 people and 15 gal. per person for additional family members.

2/ From: R. G. Werden and L. G. Spielvogel: "Part II Sizing of Service Water Heating Equipment in Commercial and Institutional Buildings," ASHRAE Transactions, Vol. 75 PII, 1969, p. iv.1.1.

* Adopted from HUD Minimum Property Standards for Solar Systems.

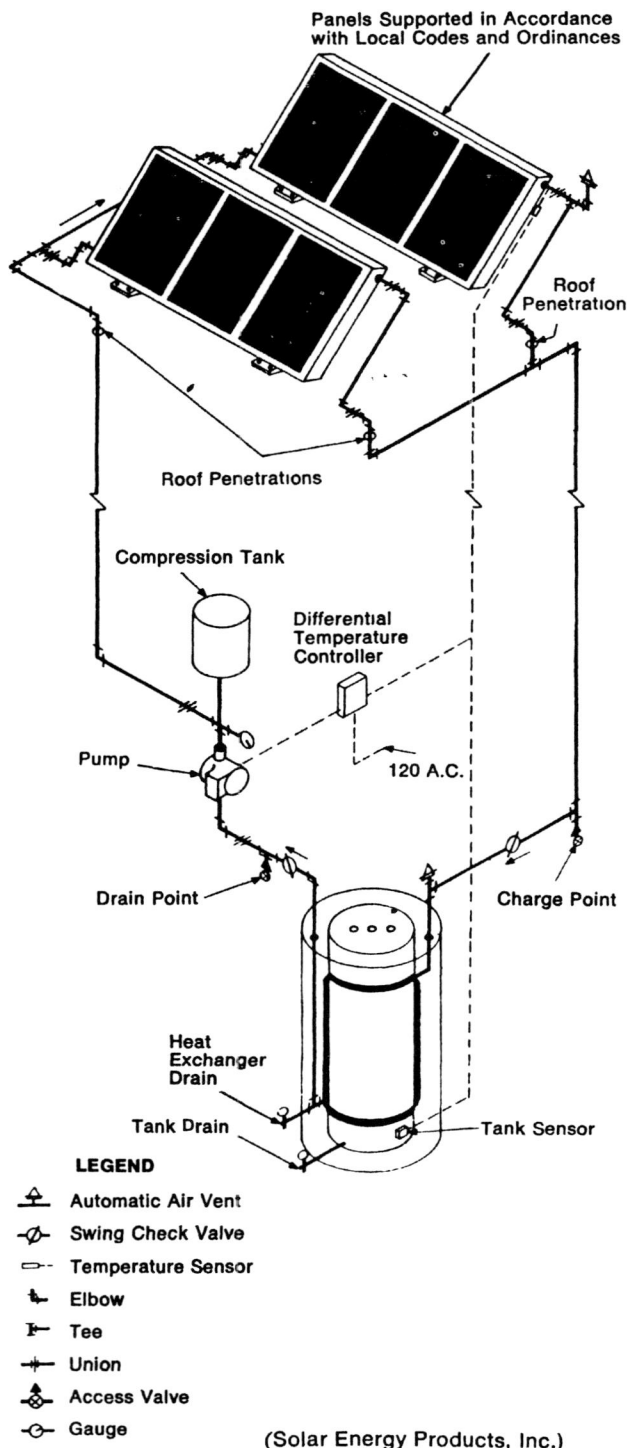

Panels Supported in Accordance with Local Codes and Ordinances

Roof Penetration

Roof Penetrations

Compression Tank

Differential Temperature Controller

Pump

120 A.C.

Drain Point

Charge Point

Heat Exchanger Drain

Tank Drain

Tank Sensor

LEGEND

⟂ Automatic Air Vent

⌀ Swing Check Valve

⊏-- Temperature Sensor

↳ Elbow

⊢ Tee

⊣⊢ Union

⊗ Access Valve

⊙ Gauge

(Solar Energy Products, Inc.)

Fig. 8-1. Typical domestic solar hot water system.

tank to the collector where it is warmed and the flow continues. Figure 8-2 schematically depicts this system. Several conditions must be maintained in order for this type of system to function. First, the tank must be located slightly above the collector, about 1-2 feet.

This permits the water to circulate through the system without the use of a pump. It also prevents reverse circulation which would otherwise occur when the collector temperature is lower than the stored water temperature. Typically, single glazed collectors and 40-80 gallon tanks are used.

This type of system is simple and relatively maintenance free; however, it possesses a number of limitations:

1. Water temperature varies depending upon the amount of sunlight available and the rate at which hot water is consumed.

2. The requirement that the storage tank be above the collector poses some design problems because the system cannot be readily fitted into all types of structures.

3. The fact that the collector always contains water means that the system in its simplest form cannot be used in climates where the temperature drops below freezing.

4. It is possible for the water in the system to exceed safe temperatures. Therefore, it is necessary that safety valves be installed to allow excess pressure to escape.

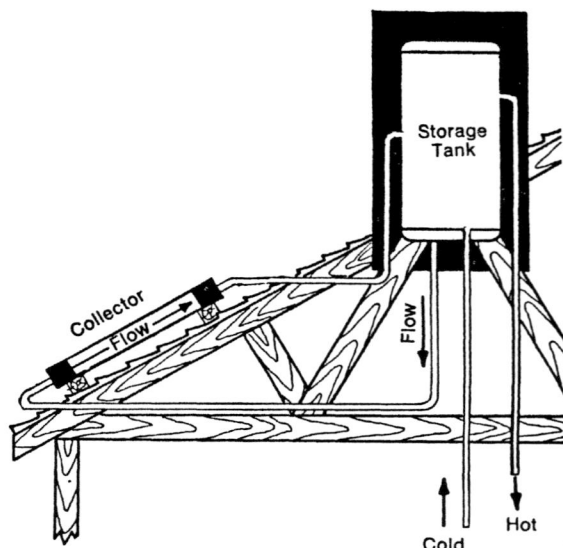

Storage Tank

Collector

Flow

Flow

Cold

Hot

Fig. 8-2. Schematic drawing of a direct heating thermo siphon circulating solar water heating system.

A direct heating/thermosiphon circulating system can be used in climates where the temperature drops below freezing, provided valves are installed which will automatically drain the collector when the temperature in the collector nears freezing. To prevent the storage tank from draining, another thermostatically activated valve between the collector and storage tank must close. Because this system of valves is somewhat complex, the possibility of malfunction makes this type of system impractical in freezing climates.

Direct Heating/Pump Circulating Systems

The direct heating-pump circulating system differs from the preceding system in that a pump is used to circulate the water through the collector.(See Figure 8-3.) Because of the installation of the pump, it is possible to locate the storage tank below the collector. For many installations, this is a much more practical arrangement because it allows the collector to be placed on the roof and the storage unit can be either on the first floor of the building or in the basement.

The difference in temperature between the collector outlet and the water at the base of the storage tank determines when the circulating pump will activate. A preset difference in temperature at the sensors of approximate 10° F causes the pump to begin pumping water through the collector. To prevent reverse flow due to the thermosiphon effect when no solar energy is being collected, a check valve must be installed in the circulation piping.

If hot water is not used at regular intervals, it is possible that the water in the collector and storage tank may exceed normal temperature. It is even possible that boiling may occur in the collector. To prevent damage to the system, a pressure/relief valve must be installed in the collector loop line entering the storage tank. This valve releases excessive pressure and permits the escape of steam.

Because this is a direct heating system, the water which is being heated is also circulated through the collector. Like the thermosiphon system, this system is subject to freezing. To prevent freeze damage to the system, it is necessary to drain the collector when freezing weather threatens. The required freeze protection is like that which was described in the previous section. The most reliable valves are mechanically driven (by springs or other mechanical devices). Therefore, in the case of a power failure, the collector

can automatically drain. Figure 8-4 indicates where the valves should be located and Figure 8-5 details a typical direct heating, drain-down installation.

The single tank system in Figure 8-5 uses an electric heater inside the tank to provide auxiliary heating of domestic water when no solar energy is available.

Because of temperature stratification within the tank (hot water at the top and colder, denser

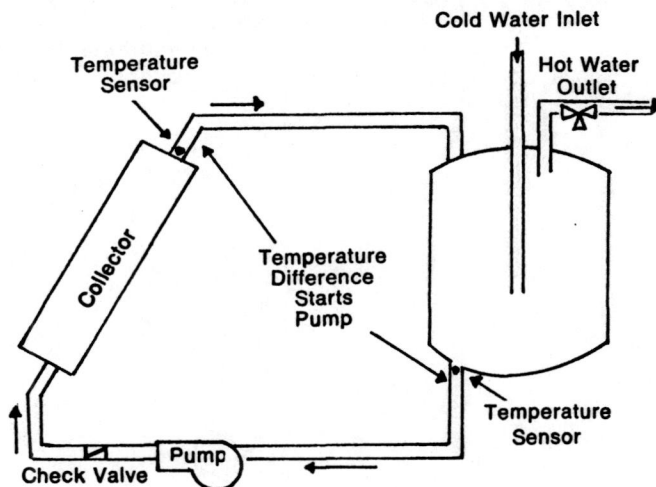

Fig. 8-3. Schematic diagram of a direct heating/pump circulating system.

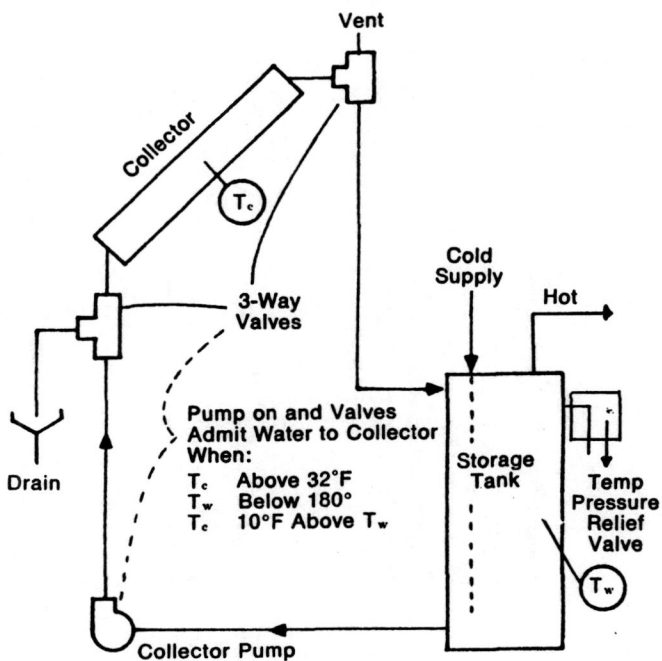

Fig. 8-4. Schematic diagram of a drain down system for a direct heating/pump circulating system.

water at bottom), the electric element is usually installed near the top of the tank. This reduces the on time of the heater and takes full advantage of the available solar energy. There is, of course, internal mixing caused by water entering and leaving the tank and this arrangement may not be as efficient as a two-tank system.

Figure 8-6 is a schematic of the system shown in Figure 8-4 but with the addition of a second tank, typically a conventional water heater.

In this arrangement, the conventional heater is connected in series with the solar preheat storage tank. The conventional heater can be gas-fired, oil-fired, or electric.

Placed in this configuration, preheated water from the solar storage tank enters the conventional heater before flowing through the hot water

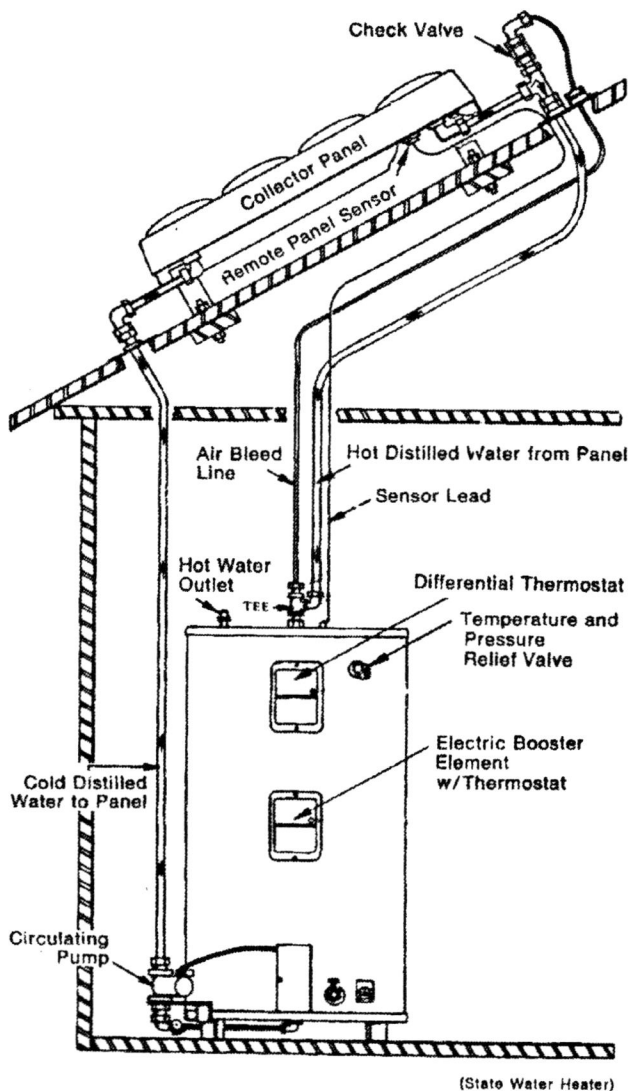

Fig. 8-6. Solar water heater in series with conventional water heater.

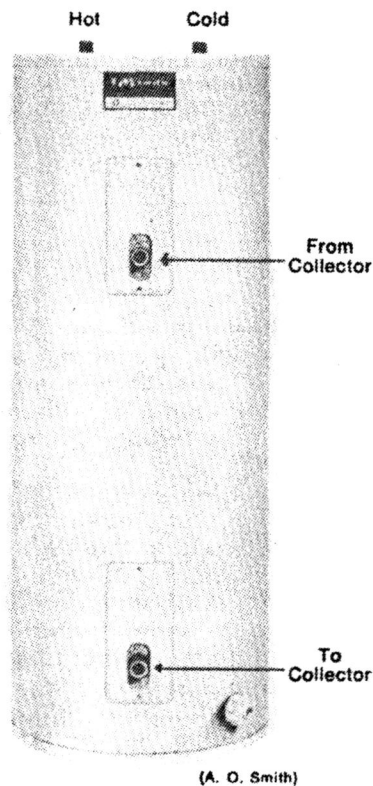

(State Water Heater)

Fig. 8-5. Preheat, storage and electric auxiliary heater in single tank.

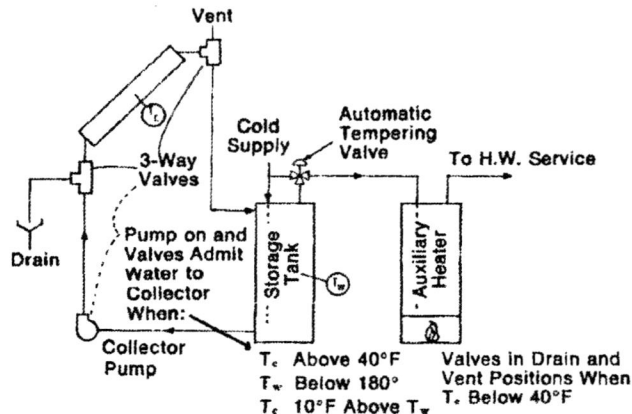

(A. O. Smith)

Solar water storage tank

service main. In this way, auxiliary heat energy is used to raise the water temperature only when solar energy is unavailable or inadequate to maintain preselected water temperature.

Indirect Heating/Circulating Systems

To overcome the problem of draining liquid collectors during periods of subfreezing weather, indirect heating solar collectors have been developed. Indirect heating systems circulate an antifreeze solution or special heat transfer fluid through the collectors. Air collectors can also be used. As a result, there is no danger of freezing and no need to drain the system.

Liquid Transfer Media. Circulating a solution of ethylene glycol and water through the collector and a heat exchanger is one means of eliminating the problems of freezing. Figure 8-7 illustrates a typical liquid system. Note that this system requires a heat exchanger and an additional pump. The heat exchanger permits the heat in the liquid circulating through the collector to be transferred to the water in the storage tank. The extra pump is required to circulate water from the storage tank through the heat exchanger. The extra pump can be eliminated if (1) the heat exchanger is located below the storage tank and (2) the pipe sizes and heat exchanger design permit thermosiphon action to circulate water from the storage tank through the heat exchanger; or (3) a heat exchanger is used that actually wraps around and contacts the storage tank and transfers heat directly through the tank wall.

Safety Devices in Liquid Media Systems. There are two major problems that might develop with liquid solar water heaters: (1) excessive hot water may enter the domestic hot water service line, and (2) high temperature-high pressure may damage collectors and storage unit.

To eliminate the first problem, a mixing valve is typically installed between the solar storage tank and conventional water heater as shown in Figure 8-8. Cold water is blended with hot water in the proper proportion to avoid excessive supply temperature. The mixing valve is sometimes also referred to as a tempering valve. Figure 8-9 details a typical connection.

To avoid excessive pressure in the collector loop carrying the antifreeze or heat transfer solution, a pressure relief valve is installed in the loop. The valve is usually set to discharge at pressure in

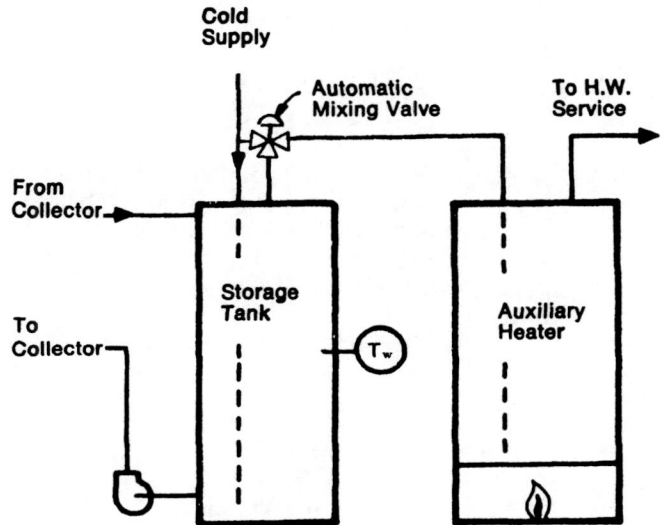

Fig. 8-8. Schematic diagram of the auxiliary heating equipment for a solar water heating system.

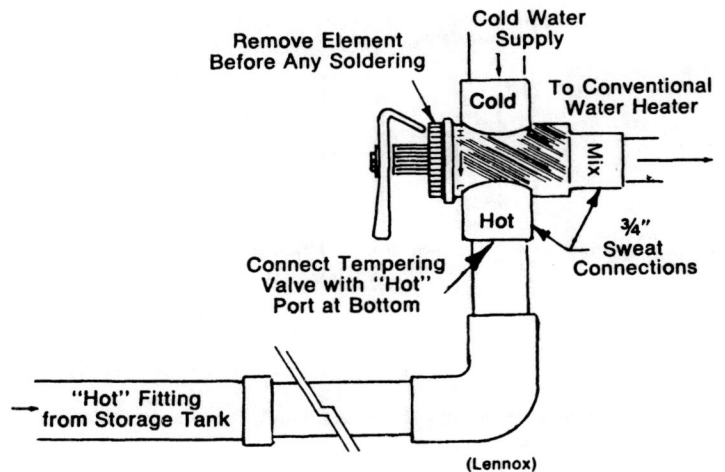

Fig. 8-9. Typical tempering valve assembly.

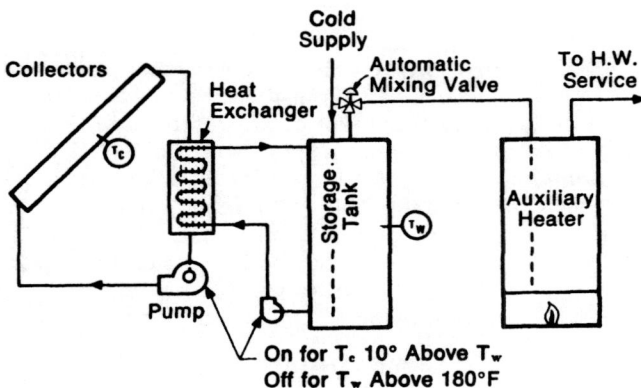

Fig. 8-7. Indirect solar water heating system.

excess of 50 psi. The relief valve is plumbed to an open drain, since fluid temperature may exceed 200° F. Remember, this is *unsafe*, contaminated fluid.

To protect the storage tank, a temperature and pressure relief valve is usually installed on the storage tank. Whenever water in the tank excesses about 210° F, the valve opens and purges the hot water in the tank. Cold water automatically enters the storage tank and provides a "heating load" for the collector loop, thereby cooling down the system. Figure 8-10 shows examples of both safety devices installed in a system.

Also shown in Figure 8-10 is the collector loop expansion tank. This device is required to "absorb" the expansion and contraction of the circulating fluid as it is heated and subsequently cooled. Any loop not vented to the atmosphere *must* be fitted with an expansion tank.

Heat Exchanger. The heat exchanger which acts as an interface between the sometimes toxic collector fluid and the potable water to be heated *must* be double-walled. That is, if one side of the heat exchanger leaks as a result of rupture or corrosion, the toxic fluid will not contaminate the water.

A conventional shell and tube heat ex-

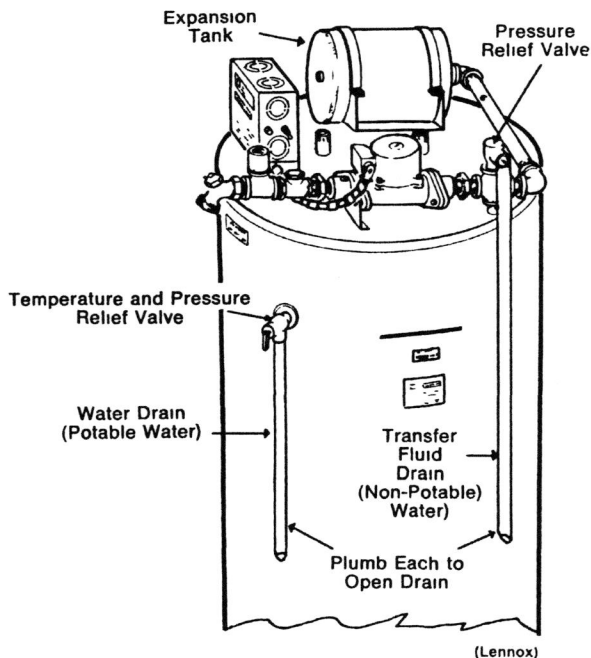

Fig. 8-10. Example of two safety valves on solar heater plus expansion tank.

changer or simple coil inside the storage tank will, most often, not meet local health code requirements.

Figures 8-11 and 8-12 detail several heat exchanger types and indicate those likely to meet most code requirements.

Air Transfer Media. Air-heating collectors can be used to heat domestic water. (See Figure 8-13.) The operation of this type system is very similar to the indirect liquid circulating system.
The primary difference is that a blower (fan) is used to circulate air through the collector and heat exchanger rather than a pump circulating a liquid.

The major *advantages* of the air transfer medium are (1) freedom from damage due to liquid leakage in the collector loop, (2) freedom from freezing and boiling, and (3) elimination of the risk of losing the expensive fluid in the collector loop. The *disadvantages,* as compared to liquid media systems, include the (1) larger piping required between collector and heat exchanger, (2) somewhat more energy required to operate the circulating fan, and (3) the need for a slightly larger collector.

Operating Cycle

To control the operation of direct or indirect water heating systems, a differential temperature controller is used to measure the temperature difference between the collector and storage, and thereby control pump operation.

Typically, when there is more than a 10° F difference between storage and collector temperatures, the pump will be started. When the temperature difference drops to less than 3° F, then the pump shuts off.

There are several modifications of this operating mode, one of which is the use of two-speed or even multi-speed pumps. Quite simply, when solar intensity is low, the pump operates on low speed; as solar radiation increases, the pump is speeded up. This is intended to improve collection efficiency.

Sizing Domestic Water Heating Systems

As with space heating, the sizing of a solar water heater system must begin with an estimate of the Btu load. Table 8-1 is taken from HUD's Minimum Property Standards for Solar Systems. The *minimum* daily hot water requirements for various residence and apartment occupancy are listed. For example, a two-bedroom home with three occupants should be provided with equipment that can provide 55 gallons per day of hot water. Many designers simply assume 20 gallons per day per person, which results in slightly higher requirements than those listed in Table 8-1.

The *second* important consideration in sizing solar domestic hot water heating systems is the required change in the temperature of the incoming water. The water supplied by a public water system usually varies from 40 to 75° F, depending on location and season of the year. A telephone call to the local water utility will provide the water supply temperature in the area. Generally, the desired supply hot water temperature is from 140° F to 160° F. Knowing these two temperatures and the volume of water required enables one to calculate the Btu requirement for domestic hot water. (See Figure 8-14.)

The required collector area to provide some portion of this Btu load can be determined in a number of ways.

First of all, a detailed FCHART performance analysis can be performed. Recall that in this procedure, a collector area is assumed for a specific locality and monthly performance (fraction of energy contributed by solar) are determined using the X and Y coordinates of the FCHART. Then, a seasonal contribution is determined.

As with whole house space heating, a number of simplified procedures have been developed.

One rule of thumb is: the amount of solar energy available at mid-latitude in the continental United States is approximately equal to 2000 Btu/

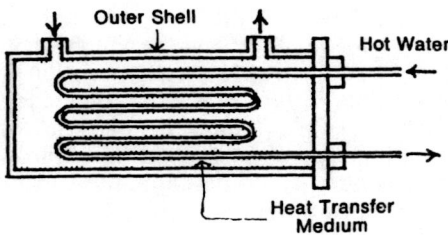

(A) Shell and Tube. This type of heat exchanger is used to transfer heat from a circulation transfer medium to another medium used in storage or in distribution. Shell and tube heat exchangers consist of an outer casing or shell surrounding a bundle of tubes. The water to be heated is normally circulated in the tubes and the hot liquid is circulated in the shell. Tubes are usually metal such as steel, copper or stainless steel. A single shell and tube heat exchanger *cannot be used* for heat transfer from a toxic liquid to potable water because double separation is not provided and the toxic liquid may enter the potable water supply in a case of tube failure.

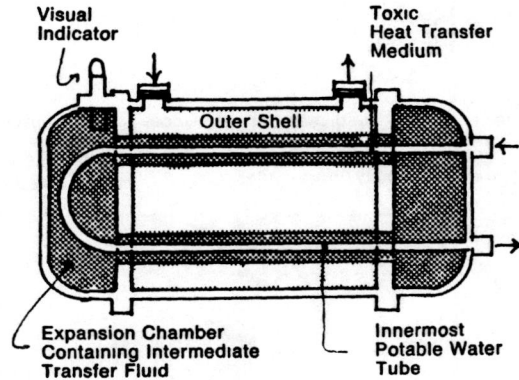

(C) Double Wall. Another method of providing a double separation between the transfer medium and the potable water supply consists of tubing or a plate coil wrapped around and bonded to a tank. The potable water is heated as it circulates through the coil or through the tank. When this method is used, the tubing coil must be adequately insulated to reduce heat losses.

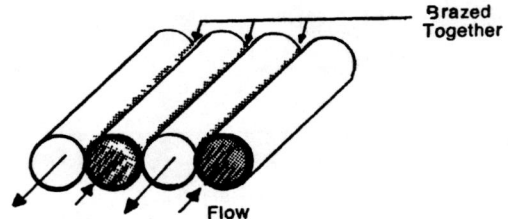

(B) Shell and Double Tube. This type of heat exchanger is similar to the previous one except that a secondary chamber is located within the shell to surround the potable water tube. The heated toxic liquid then circulates inside the shell but around this second tube. An intermediary non-toxic heat transfer liquid is then located between the two tube circuits. As the toxic heat transfer medium circulates through the shell, the intermediary liquid is heated which in turn heats the potable water supply circulating through the innermost tube. This heat exchanger can be equipped with a sight glass to detect leaks by a change in color — toxic liquid often contains a dye — or by a change in the liquid level in the intermediary chamber, which would indicate a failure in either the outer shell or intermediary tube lining.

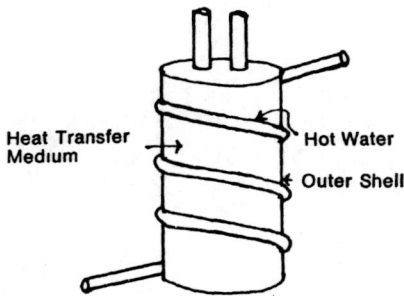

(D) Parallel Tube Heat Exchanger. A double walled exchanger suitable for domestic water heating using toxic fluid in collector loop.

Fig. 8-11. Examples of heat exchanger designs.

square foot/day. Assuming a collector efficiency of 40 percent, this means that 800 Btu/square foot/day can be collected with a properly installed collector. Using the example in Figure 8-14, the collector should contain approximately 137 square feet (109,956 ÷ 800 = 137), assuming 100 percent satisfaction of hot water energy needs. If a collector of this size were used, however, the higher summer radiation levels and warmer tem-

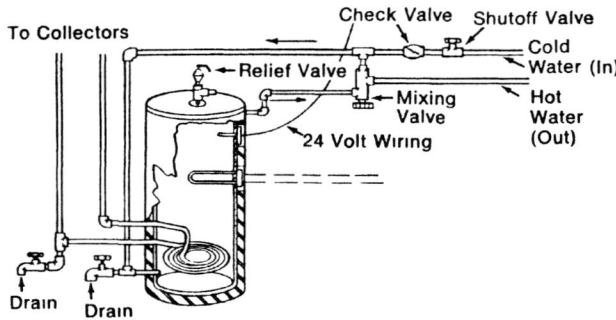

Coil inside the tank (left) may not meet local code requirements if toxic fluid is used in the collector loop. The addition of a shell and tube heat exchanger between collectors and coil would probably satisfy most codes.

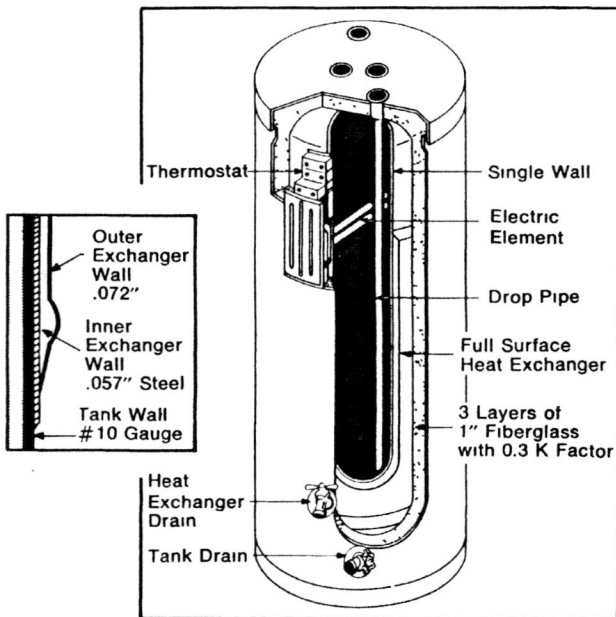

(Solar Energy Products)

Tank at right has "wrap around" heat exchanger through which heated fluid from the collectors flow. Heat is transferred across a "double wall" comprising one side of heat exchanger and wall of storage tank. This meets HUD minimum property standards.

Fig. 8-12. Coil inside the storage tank and wrap around heat exchangers.

peratures would result in excess capacity most of the time. A more practical approach is to provide nearly 100 percent solar hot water in July which might then average out to 70 percent contribution for the year. Thus, a collector area of 0.7 x 137 or about 96 square feet might be a more realistic installation.

The "rule-of-thumb" sizing procedures outlined above assume that the collector is installed facing due south and inclined at an angle equal to the local latitude plus 10 degrees. Modification of these optimum collector installation procedures will reduce the effectiveness of the collector. In case the ideal installation cannot be achieved, it will be necessary to increase the size of the collector to compensate for the loss in effectiveness. The procedures for computing the additional collector area are identical to those described in Lesson Six. Therefore, if it is desirable to install a domestic hot water heating system which faces southwest or southeast, refer to the section in Lesson Six entitled "Effect of Facing the Collector East or West of Due South." Likewise, if the collector must be tilted at an angle different from latitude plus 10 degrees, the procedure in Lesson Six entitled "Effect of Installing Solar Collectors at Angles Other than Optimum" should be used to compute the size of the collector. However, a closer approximation can be obtained through the use of Figure 8-15.

The vertical scale (F) is the fraction of the annual water heating load satisfied by solar energy. The horizontal scale is AS/L, where A is the collector area, S is the average *daily* solar radiation on a horizontal surface in January (from Appendix A), and L is the *daily* water heating load in January.

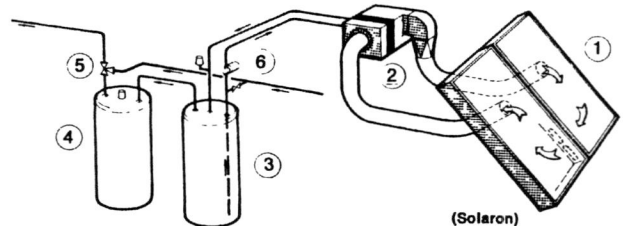

1. Air collectors
2. Air to water heat exchanger with blower
3. Preheat storage tank
4. Conventional water heater
5. Mixing valve (may be required)
6. Heat exchanger to storage pump

Fig. 8-13. Schematic diagram of an air transfer medium solar water heating system.

Obviously, a system designed to satisfy the total or near total Btu load in winter would be much oversized in summer. Why? Because the domestic water heating load would change but slightly while the amount of collectable solar radiation would be substantially larger—perhaps nearly double.

Observations have indicated that a collector sized to provide about 60 percent of the January hot water load will provide nearly all the hot water energy needs in June.

SMACNA Installation Standards indicate that a solar water heating system should be designed to provide no less than 50 percent of the *annual* energy needs.

With these points in mind, Figure 8-15 can be used to estimate a reasonable collector area by, first, assuming the desired annual contribution and, then, determining the value of AS/L from the chart. Values of F from 50 to 70 percent appear to be reasonable assumptions. An example of the application of this procedure is given in Figure 8-16.

Example Problem:

Given that a family of six people live in a home where the incoming water temperature is 40°F. and the requirement is for 150°F hot water, calculate the BTU requirement

1. Compute: Water requirements = Number of people × 20 gallons per day
   = 6 × 20
   = 120 gallons of hot water used per day

   Because a gallon of water weighs 8 33 pounds, the BTU requirement per day for hot water can be found by:

2. Compute: Heat Required = gallons per day × 8.33 × temperature rise
   = (120 × 8.33) (150 − 40)
   = 109956 BTU per day

Fig. 8-14. Calculating the BTU requirements for heating domestic hot water.

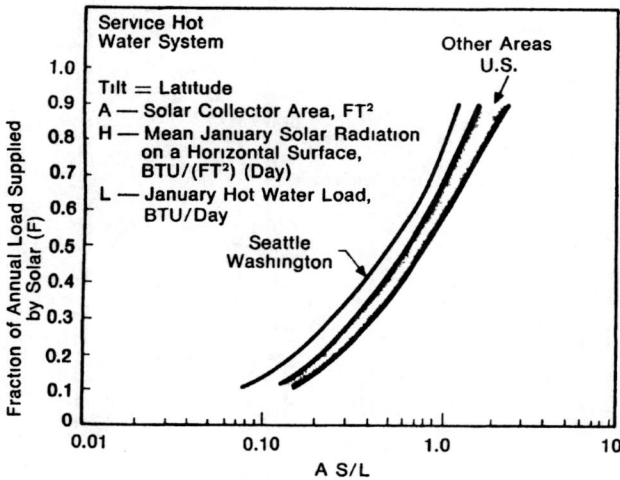

Fig. 8-15. Fraction of annual load supplied by solar as a function of January conditions (water system).

Suppose a family in Boston, Mass., desired to install a solar water heating system. Using figure 8-14 we find that a minimum of 70 gallons of hot water are required per day. From the local water utility we learn that January service water is supplied at 40°F The design hot water service is set at 140°F. Thus, we can calculate the BTU requirement

$$L = W \times 8.33 \times (t_2 - t_1)$$

Where

   W = Daily hot water requirement (gallons)
   t_2 = Hot water service temperature
   t_1 = Temperature of cold water supply

Thus

$$L = 70 \times 8.33 \times (140 - 40)$$
$$= 64,141 \text{ Btu's/day}$$

From Appendix A, we find that the average daily January solar radiation is 506 Btu / sq ft / day Thus

$$S = 506 \text{ Btu / sq. ft. / day}$$

From figure 8-16, the mean value for A S/L for a 60% (F) is 0 9 To determine the size of the collector we use the ratio

$$\frac{AS}{L} = 0 9$$

Substituting from above

$$\frac{A \times 506}{64,141} = 0 9$$

Solving for A

$$A = \frac{64,141 \times 0 9}{506} = 114 \text{ sq ft of collector}$$

Please note that the curves of Fig. 8-16 assume that the collector is tilted at an angle equal to the local latitude. To compensate for any collector tilt refer back to lesson 6.

Fig. 8-16. Determining the size of the collector for a solar hot water heater.

Sizing Other Components

For the most part, solar domestic water heating systems are available as prepackaged components. Figure 8-17 illustrates two examples of prepackaged domestic solar water heaters. These eliminate the need to size the storage tank, expansion tank, and pump. For an individual wishing to select individual components, it will be necessary to make the same type of calculations as for whole-house heating to determine the sizes of such components. Tank storage would typically be based on one day's supply of energy based on the daily Btu load.

Figure 8-18 illustrates a typical piping and wiring arrangement for a solar water heating system.

Economic Considerations

As with space heating, the prospective buyer of a solar assisted domestic water system will be interested in the savings accrued for the added investment required for a solar system. Most manufacturers of packaged solar water heating systems provide some type of economic analysis to assist the installing contractors in selling their customers. For example, payback time for fuel savings to equal total investment may be as little as six to nine years at the present time.

As with space heating, there is a computer service called SOLCOST that a contractor can use to provide a complete analysis based on the infor-

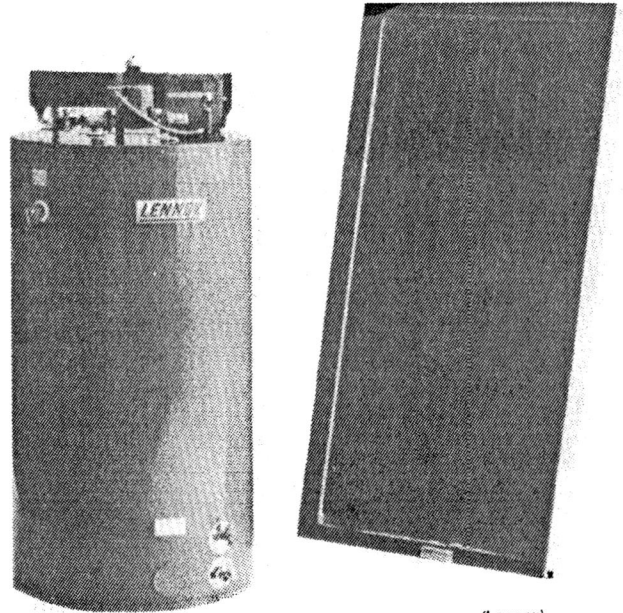

(Lennox)

(Taco)

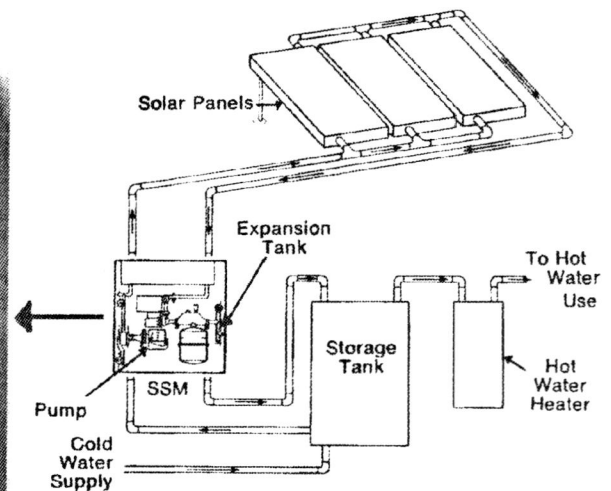

Fig. 8-17. Examples of prepackaged solar assisted domestic water heaters.

8-10

mation supplied to the computer service. Appendix E includes a sample of the SOLCOST Solar Service Hot Water Worksheet. Figure 8-19 illustrates a typical economic "print out" for a water heating system.

This service also includes collector size optimization calculation that will provide the customer the optimum savings over the life of the equipment.

For full details contact: International Business Services, Inc., Solar Group, 1010 Vermont Avenue, Washington, D.C. 20005, (202) 628-1450.

Swimming Pools with Solar Assist

Solar collectors can be used to heat swimming pools that are both indoors and out-of-doors.

Collectors used to heat outdoor pools are usually less elaborate than space heating collectors simply because they are used in the spring and summer months. Thus, single glazed or even unglazed absorber panels can be used.

Many unglazed panels are made of plastic or rubber as well as metal. However, in the case of metal, swimming pool water is never circulated

through all-aluminum panels because of excessive corrosion.

Figure 8-20 illustrates both a manual system and an automatic solar assisted pool heating system.

Under natural conditions, pool water will gain some heat during the day and then lose some heat at night. Typically, pool water may exceed the average ambient temperature by up to 10 degrees under these conditions.

Outdoor swimming pools lose heat in several ways, namely by:

1. conduction through the pool material to the ground.

2. convection at the pool surface.

3. evaporation at the pool surface.

4. radiation at the pool surface.

Conduction losses are usually small relative to the other modes of heat loss. Thus, to make full use of any solar assist (natural or with panels), outdoor pools should be provided with pool covers so that convective, evaporative, and radiation losses are minimized.

One type of pool cover of particular interest to the solar technician is the *transparent* cover. This product allows considerable solar energy to pass through during the day to help heat the pool water as well as minimize pool heat losses at night.

Considering typical pool heat losses, the usual rule-of-thumb sizing procedure suggests that, for adequate performance, the collector area should equal half the pool area for south facing collectors. For other directions, additional area is required. Figure 8-21 lists a typical recommendation for several orientations in a mild climate application.

Proper tilt depends on the season of the year for which pool heating is desired. In northern climates, pool heating is generally required in the summer, thus the preferred tilt is latitude minus 10 degrees. In southern areas, winter pool heating might be the objective. In this case, the preferred tilt is latitude plus 10 degrees.

Most manufacturers of collectors intended for pool applications offer design assistance and make recommendations on control of the system. For additional information, contact the National Swimming Pool Institute, 2000 K Street, N.W., Washington, D.C. 20006.

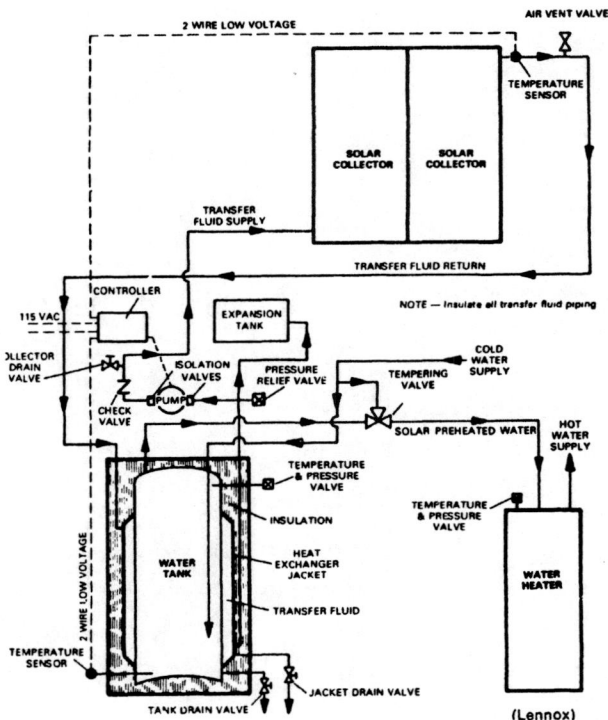

Fig. 8-18. Piping and wiring arrangement for indirect heating with storage.

Solar Components Sizing

Solar storage tank size for this collector is **84** gallons.
Heat Exchanger: None
Pump Characteristics: 1/20 horsepower, 1.40 gallons per minute.
Pipe Diameter (Flow Area): ¾ inch

Collector Size Optimization by SOLCOST

Collector Type — Flat Plate 1 Glass Selective
Best Solar Collector Size for Tilt Angle of 43 Degrees is 56 Sq. Ft.
Solar System Costs (dollars) — 400.00 fixed   700.00 Collector   105.00 Storage
Installation Costs (dollars) — 680.00

Cash Flow Summary

Yr.	(A) Fuel/Utility Savings	(B) Maint. + Insur.	(C) Property Tax	(D) Annual Interest	(E) Tax Savings	(F) Loan Payment	(G) Net Cash Flow	
1	163.	29.	48.	169.	72.	363.	-205.	-384. (Down Payment)
2	174.	30.	50.	148.	65.	363.	-203.	
3	186.	31.	52.	124.	58	363.	-201.	
4	199.	32.	54.	98.	50.	363.	-200.	
5	213.	34.	56.	68.	41.	363.	-198.	
6	228.	35.	58.	36.	31.	363.	-197.	
7	244.	36.	61.	0.	20.	0.	167.	
8	261.	38.	63.	0.	21.	0.	181.	
9	279.	39.	66.	0.	22.	0.	196.	
10	299.	41.	68.	0.	23.	0.	212.	
11	320.	43.	71.	0.	23.	0.	229.	
12	342.	44.	74.	0.	24.	0.	248.	
13	366.	46.	77.	0.	25.	0.	268.	
14	392.	48.	80.	0.	26.	0.	290.	
15	419.	50.	83.	0.	27.	0.	313.	
16	448.	52.	87.	0.	29.	0	339.	
17	480.	54.	90.	0.	30.	0.	366.	
18	513.	56.	94.	0.	31.	0.	395.	
19	549.	58.	97.	0.	32.	0.	426.	
20	588.	61.	101.	0.	33.	0.	459.	
Totals	6663.	857.	1430.	643.	683.	2178.	2885.	

Payback time for fuel savings to equal total investment — 8.9 years
Payback time for net cash flow to offset down payment — 13.3 years
Rate of return on net cash flow — 8.5 percent
Annual portion of load provided by solar — 35.3 percent
Annual energy savings with solar system — 11.6 million BTUs

Tax Savings = Income Tax Rate $\times$ (C + D)
Net Cash Flow = A - B - C + E - F

Fig. 8-19. Example of "Solcost" analysis.

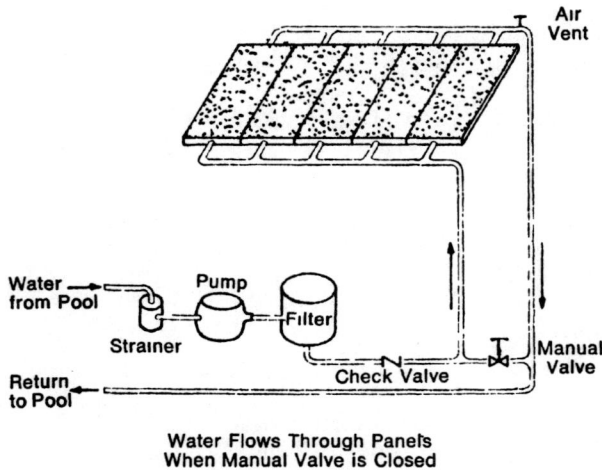

Water Flows Through Panels
When Manual Valve is Closed

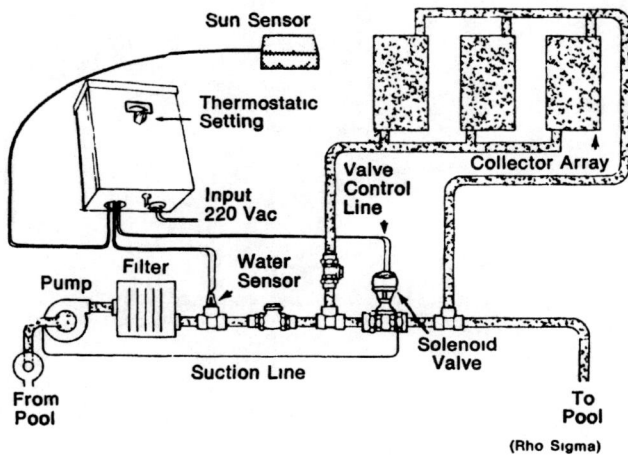

(Rho Sigma)

Fig. 8-20. Manual and automatic solar pool heaters.

1 Determine the surface area of the pool.	
2. Select appropriate proportion as determined by collector installation.	

.50 - .70	.70 - .80
Panels facing south with tilt of latitude + 10°	Panels facing southwest or southeast **or** Panels not tilted at latitude + 10°

3. Multiply.
Surface Area × Proportion
to determine the
collector area required

Fig. 8-21. Estimating the size of the collector for a swimming pool installation.

SUMMARY

The three basic types of solar domestic hot water heating systems have been described:

1. Direct heating/thermosiphon

2. Direct heating/pump circulating

3. Indirect heating/pump circulating

Transfer mediums and heat exchangers for these systems were discussed. The various operating cycles and controls were considered along with the required safety device.

Sizing of the various components of domestic hot water heating systems were described in relationship to the economic factors which affect the decision to purchase such a system. Also, some consumer information regarding the use of solar systems to heat swimming pools was provided. Finally, suggestions for obtaining assistance in preparing sizing and cost analyses for solar heated DHW and swimming pools were presented.

HEATING SYSTEM INSTALLATION

Many factors influence decisions about the feasibility of installing a solar heating system. Previous lessons have dealt with the problems of solar insulation dictated by the geographic location of the structure. Problems of air-versus-liquid collector loops have been discussed. Various control mechanisms have been presented. The economic factors of solar heating systems have been identified. Total system operations were explained with the inclusion of domestic hot water (DHW) requirements.

At this point, decisions must be made about aesthetic values and physical space requirements of the solar heated structure. How do collectors affect the outward appearance of the building? How can exterior, above-ground equipment be hidden? Where can the heat storage (Figure 9-1), transfer system, and controls be placed? How noisy will the system be when it is fully operational? In addition to all of the space requirements for the solar heating equipment, where can

the 100 percent auxiliary heating system for the occupants' comfort and hot water needs be located?

These questions all affect the utilization of space on a building site. Where is the least valuable space on the property? What area will the occupant have to relinquish for the various pieces of equipment? Can an area called an "Equipment Room" be accommodated? Such decisions as these must be made in the design phase of the solar heating project, regardless of whether the installation is in a new structure or a *retrofit* in an existing building.

Guidelines will have to be obtained from the component manufacturers so that inspection, maintenance, and repairs can be done. For example, how large will the attic access or the equipment room door need to be to accommodate component (pump, blower, etc.) removal, service, and installation?

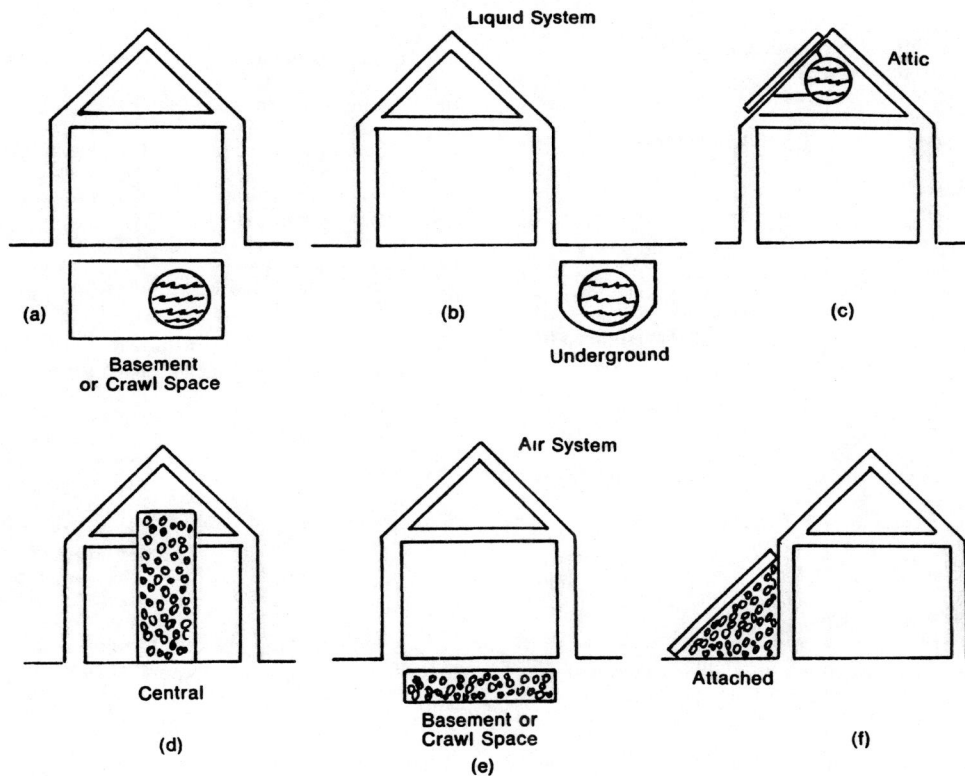

Fig. 9-1. Possible heat storage unit locations.

Scheduling is ready to begin when these decisions are finalized in contractual documents between the occupant and installer. The occupant should have information about what is being purchased and installed. What percentage of the total heat load demand and hot water needs are being provided by the solar unit? What are the operating requirements (utilities, maintenance, and supplies)? How long can the system be expected to last? Answers to these questions are forthcoming.

New Construction Installation

This section of the lesson deals with new construction and is divided into two parts according to the fluid medium (liquid or air) being used. Components and systems will be discussed as they would normally occur in scheduling construction of the entire structure. Appropriate reference to SMACNA installation standards, where noted, should be read. (See Tables 9-1 and 9-2.)

Solar System Installation Scheduling
(Liquid Heating System)

Table 9-1

Solar Component Installation	Building Construction Stage	Solar Component Installation	Building Construction Stage
I. Thermal Storage Container		3. Insulation and inspection of piping	*Prior* to drywall
1. Structural base	*During* foundation, concrete work	4. Installation of auxiliary furnace and/or cooling unit	Same as for conventional system
2. Placement of container	*Following* foundation, concrete work	**IV. Domestic Hot Water System Installation**	
3. Piping connections	*During* rough-in plumbing phase	1. Installation of heat exchanger and piping to preheat	*Prior* to insulation and drywall
4. Leak test filling (Note: May require subsequent draining to prevent freezing)	*During* rough-in plumbing phase leak test	2. Leak test of piping	*During* rough-in plumbing leak test
5. Insulation	*Following* leak test	3. Insulation and inspection of piping	*Prior* to drywall
II. Solar Collector Subsystem		4. Installation of preheat and auxiliary DHW tanks	Same as for conventional DHW systems
1. Preparation of support structure	*During* framing phase	**V. Controls**	
2. Placement on roof or support structure	*Prior* to insulation, drywalling, and plumbing stack-out. Also prior to finishing roof and roof flashing.	1. Sensors	Each sensor should be installed with each component it represents. For example, the storage tank temperature sensor should be installed during the fabrication of piping connections to the storage unit, the collectgor absorber temperature sensor during the collector installation, etc.
3. Installation of piping to equipment room	*Prior* to insulation and drywall		
4. Leak test of collector loop	*Prior* to insulation and drywall		
5. Insulation and inspection of collector piping	*Prior* to drywall		
III. Heat Transfer Subsystem		2. Control Panel	Same as for conventional thermostat (Note: Early installation can provide solar heating during construction)
1. Installation of piping, pumps, valves, etc	*Prior* to insulation and drywall (Note. Early installation could provide solar heating for drywall work)		
2. Leak test of plumbing	*During* rough-in plumbing leak test		

Liquid Systems

The first activity in constructing the solar heating system involves the heat storage unit. The tank manufacturer's literature will provide data concerning the locations of inlets and outlets, sensor fittings, and tank-anchoring methods. (SMACNA sections 11.1, 11.2, 11.3, and 11.4.)

Heat Storage Unit. Heat storage units for space heating are manufactured of steel, fiberglass, or concrete in a factory and delivered to the building site for installation. A butyl rubber lined concrete block system erected on-site is another alternative. Footings must be provided (a four-inch concrete slab is normally adequate). Footings for the containers to hold liquids, especially for heat storage and possibly domestic hot water storage, should be dug and poured when the basement or regular building foundation is under construction. Anchor bolts to secure the tanks can be placed at this time. Pouring footings for surface or underground tanks outside the building should be considered at this time for scheduling efficiency. The size of the footings will have to be engineered on the basis of the total tank and liquid weight when filled. Footings must be high enough so that there will be about 6 inches of space under the tank after pouring the basement floor to allow for tank insulation. (See Figure 9-2.)

Solar System Installation Scheduling
(Air Heating System)
Table 9-2

Solar Component Installation	Building Construction Stage	Solar Component Installation	Building Construction Stage
I. Pebble-Bed Storage Unit		3. Installation and inspection of duct insulation	*Prior* to drywall
1. Structural base	*During* foundation, concrete work	4 Installation of auxiliary furnace	Same as conventional system
2. Fabrication of container	*During* or immediately after concrete, foundation work	**IV. Domestic Hot Water System Installation**	
3. Rock filling of container	*Prior* to roofing over of basement or space containing storage unit	1 Installation of heat exchanger and piping to preheat	*Prior* to insulation and drywall
4. Ducting connections	*During* heating distribution system ductwork	2 Leak test of piping	*During* rough-in plumbing leak test
5. Insulation	*During* fabrication of unit	3 Insulation and inspection of piping	*Prior* to drywall
II. Solar Collector Subsystem		4. Installation of preheat and auxiliary DHW tanks	Same as for conventional DHW tank
1. Preparation of support structure	*During* framing phase	**V. Controls**	
2. Placement on roof or support structure	*Prior* to insulation, drywalling, or plumbing stackout. Also prior to finishing roof and roof flashing	1 Sensors	Each sensor should be installed with the specific component. For example, the pebble-bed storage temperature sensor should be installed during the filling of the pebble-bed storage container, collector absorber temperature sensor with the collector installation, etc
3. Installation of ducting to equipment room	*Prior* to insulation and drywall		
4. Leak test of collector loop	*Prior* to insulation and drywall		
5. Insulation and inspection of collector ducting	*Prior* to drywall		
III. Heat Transfer Subsystem		2. Central Control Panel	Same as for conventional thermostat (Note: Early installation can provide for solar heating during construction)
1. Installation of ductwork, dampers, blowers, etc.	*During* normal sheet metal work (Note: Early installation could provide solar heating for drywall work)		
2 Leak test of ducting	*Prior* to drywall and insulation		

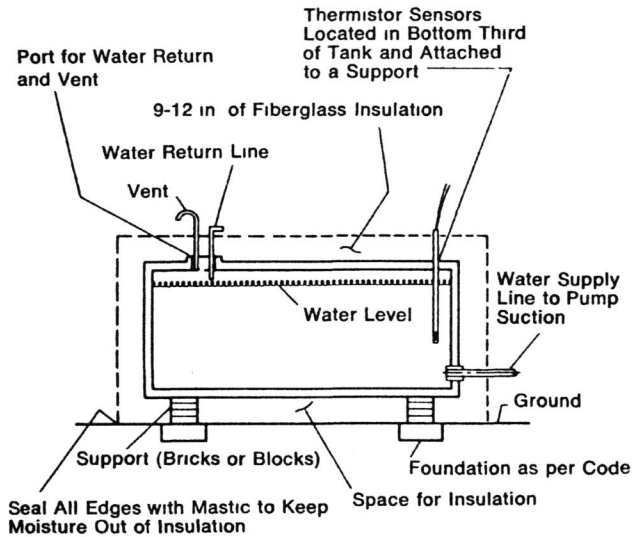

Fig. 9-2. Typical non pressurized above ground installation.

NOTE: The installer should be cautioned that underground installation of concrete storage tanks will present the following difficulties:

(1) The thermal insulation will be more difficult to install and seal from moisture.

(2) The tank and pump installation will be more difficult. A hole must be dug for the tank, and the pump would have to be installed in a dry well.

(3) Maintenance or repairs would be more difficult.

Whether the storage tank is installed indoors or outside, it should be insulated with a low-thermal-conductivity type insulation, such as 9-12 inch-thick fiberglass. If the storage tank is installed outside, the insulation must be waterproofed with roofing cement or enclosed in a tool shed or other weatherproof shelter.

Once the basement walls are erected, the tank can be placed. This operation will probably require a crane. Care must be taken to plug all openings to prevent contaminants from entering the tank. Protection against inadvertent damage while the building is under construction should be provided for the tank, especially if it is glass or stone-lined. Impact dents may fracture the lining and greatly reduce the life expectancy of the storage tanks. Check all glass and plastic inside surfaces as well as the outside galvanized coating for imperfections and delivery damage before placing the unit. If fiberglass tanks are used, they *must* be rated to withstand design storage temperatures to 180° F.

Preparations for Collector Installation. Once the tank(s) are secured and protected, construction on the structure can continue as normally scheduled: steel "I" beams can be placed; floor joists can be installed; sub-floorings can be laid; and exterior walls can be framed, sheathed, and erected.

Roof framing is next. During the design phase, a decision must be made to either make the collectors an integral part of the roof or to have them mounted above the roof as in Figure 9-5. Three possible mounting design arrangements are shown in Figure 9-6. Actual solutions to these designs approaches vary with the component manufacturer. A few examples are illustrated in Figures 9-7 through 9-9. Collector manufacturer should provide complete details on mounting collectors.

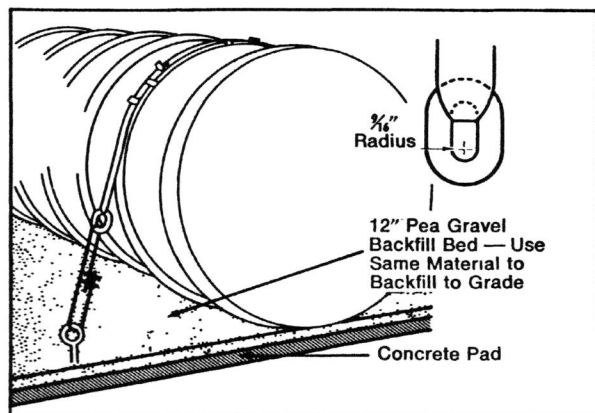

Fig. 9-3. Turnbuckle hold down straps for underground placement of fiberglass storage tank.

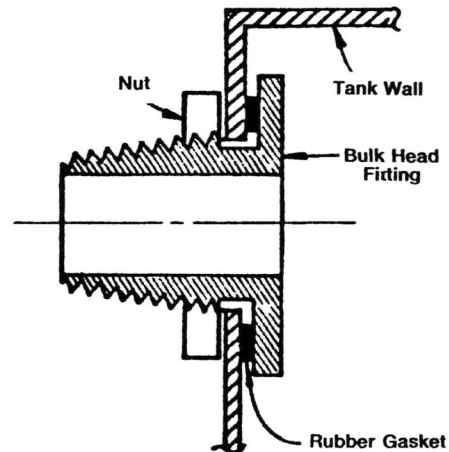

Fig. 9-4. When field connections must be made into a tank a bulk head type connection should be used.

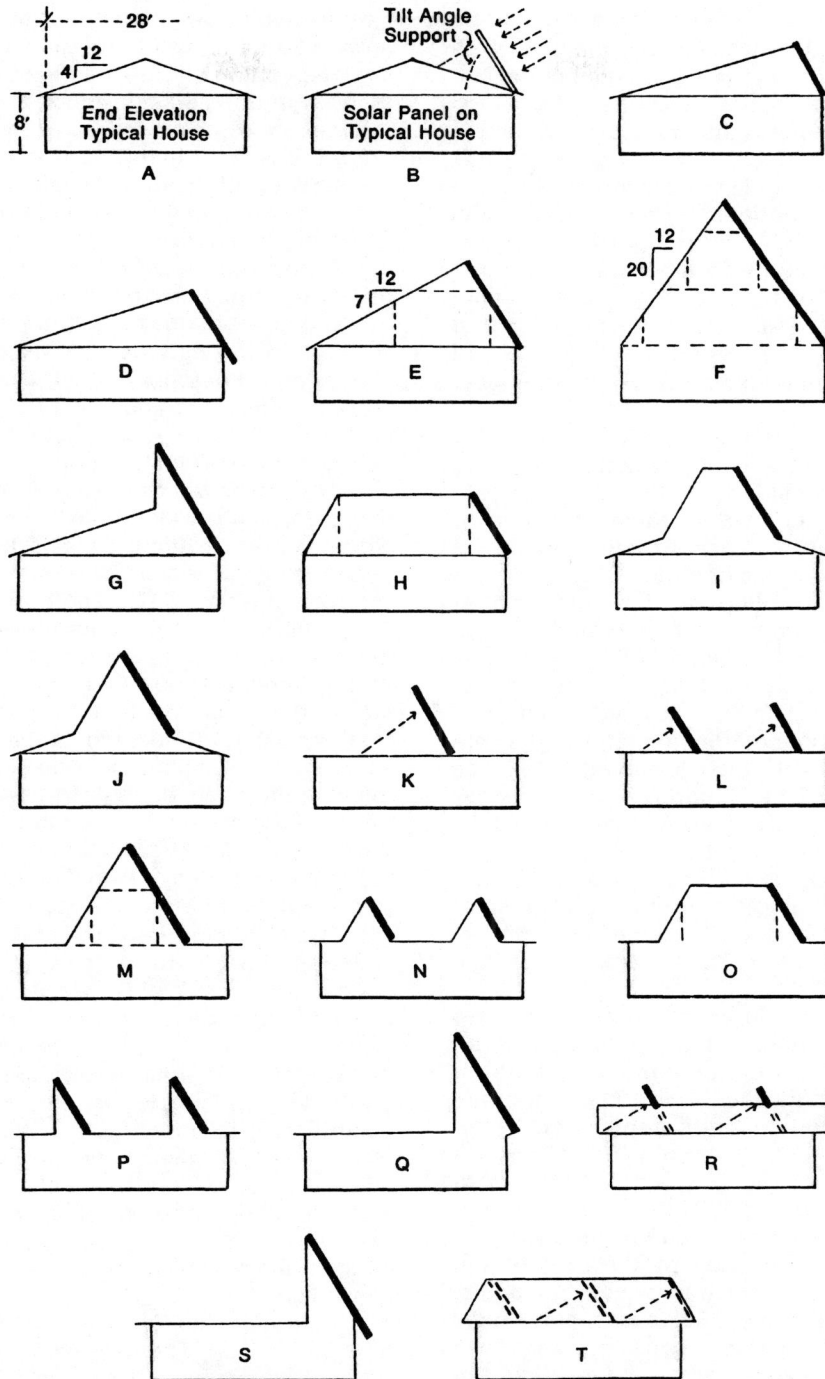

Fig. 9-5. Collector/roof configurations.

Elevations B, C, D, K and L are probably the most economical methods of accommodating a steep tilt angle for solar collectors

Elevations E, H, M and O are probably the next most economical, recognizing that they provide additional living area on the second floor.

Elevations I, J, M and possibly O, depending on proportions, must be treated carefully.

Elevations G, N, P, Q and S are possibilities but are more costly than the first group and especially more costly if clearstory space and/or fenestration are to be provided.

Elevations R and T are of some interest in that they provide a method of hiding the collectors completely, but their comparative cost needs special analysis.

Elevation F is obviously very expensive and only applicable under certain conditions. It does provide a three-story opportunity.

(NAHB

In any case, the trusses or rafters should be set with consideration for securing the collectors in place and supporting the total weight of the collector array. The possibility of damage from high winds and/or snow weight must also influence any roof framing decisions. There is almost always a problem of roof supports interfering with inlet-manifold and outlet-manifold piping at the eaves and ridge. This may require lowering the collector array. Keep in mind an important solar heating system design criterion, that piping should have as few constrictions and directional changes (elbows, etc.) as possible to keep head pressures to a minimum, and to assure complete drain down.

Interior wall partitions should be framed-in either prior to, during, or immediately following roof framing. Flexibility in scheduling this activity is determined by the need for these partitions to be load-bearing. After these structural members are in place, the roof may be installed.

The steps in roofing the structure involve sheathing, laying down roof felt, and cutting openings for piping. Roofing and flashing will be installed according to the design decision about collector panels being part of the roof structure or whether they should be attached later. If they are added later, the panel supports should be in place when the roofing is applied. This support may need to extend down to, and rest on, a typically non-supporting interior partition.

Collector Array Installation. (SMACNA Section 5-3, Item D) Collectors are the heart of solar heating. Determining their placement is probably the most critical of any of the activities in the construction project.

First, the tilt angle at which collectors are mounted is a key factor affecting the absorption of solar energy. A good rule-of-thumb is that, for heating, the tilt angle should be 15° greater than the latitude designation for the geographical location of the structure. A location of 40° North latitude will have collector arrays set at approximately 55° from horizontal. (Precise tilt is not critical, however, as mentioned previously.)

The architect may design the roof of the structure to the prescribed collector tilt angle. This procedure provides for the most rigidly mounted collector. It also means that snow will probably not interfere with solar radiation reaching the cover of the collector because of the steep angle. Most collectors (but not all) are installed with flow channels in the *vertical* position.

Collector arrays should be installed facing south in the northern hemisphere. This is neces-sary to obtain the maximum solar radiation. The array may be on the front, side, or back of buildings depending on the direction that the structure faces. Plumbing stacks, vents, and other roof protrusions must be routed away from the collector area. For direct mounting, the collectors should be installed after the roof felt has been laid, or after the asphalt shingles are in place if stand-off mounting is planned.

Collectors *must* be inspected when they arrive on the site. There cannot be any bent casings, cracked cover plates (a common problem), imperfections in the surface coating on the absorber plate, nor improperly installed rubber plate mounts. These problems will reduce the efficiency of the collector. Also, check for missing parts before beginning installation.

Collectors will need to be hoisted to the roof. (Panels typically weigh from 150 to 175 pounds so, without lifting equipment, this is at least a two person job). Collectors should be secured when they are in place. Then, additional collectors are lifted and placed next to (or above) the other collectors with spacing according to the manufacturer's specifications to allow for metal expansion and contraction. Installers should be certain that all flashing is secured as the installation progresses. The flashing must be kept from interfering with the removal of glass cover plates so that the flashing will not be disturbed if a glass is broken. Many manufacturers also recommend that collector faces be covered during assembly to avoid high temperatures inside "dry" collectors.

The temperature of the collector array must be monitored. Figure 9-10 shows a thermistor sensor mounted on the absorber plate. It should be at the outlet of the last collector in the array. An alternative is to attach the sensor to the inlet or outlet piping. In either case, the sensor is placed on the last unit where the temperature is highest according to the manufacturer's specifications. It is important that sensor lead-in wires be placed in conduit leading from the roof to the differential thermostat on the control panel to protect the control system from lightening. Usually 18 to 20 gauge wire can be used to connect sensors to controllers.

Rough-In Plumbing. (SMACNA Sections 8.1, 8.2, 8.3, and 8.4) The next series of events in the construction schedule involving the solar heating system is rough-in plumbing. During this phase of construction, the inlet and outlet collector manifolds are fabricated by soldering and/or threading the pipe fittings into the collector fittings. Figure 9-11 illustrates two of several connecting techni-

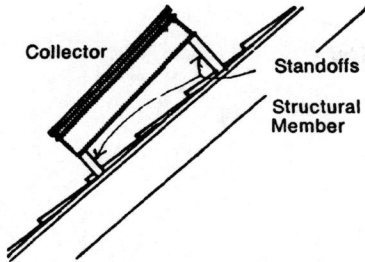

1. Rack Mounting — Collectors can be mounted at the prescribed angle on a structural frame located on the ground or attached to the building. The structural connection between the collector and the frame and the frame and the building or site must be adequate to resist any impact loads such as wind.

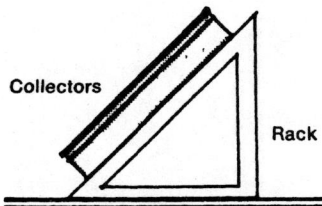

2. Stand-Off Mounting — Elements that separate the collector from the finished roof surface are known as stand-offs. They allow air and rain water to pass under the collector thus minimizing problems of mildew and leakage. The stand-offs must also have adequate structural properties. Stand-offs are often used to support collectors at an angle other than that of the roof to optimize collector tilt.

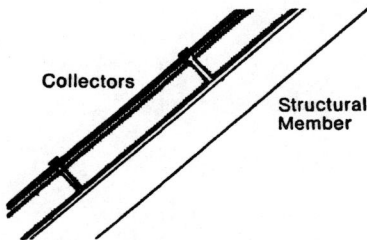

3. Direct Mounting — Collectors can be mounted directly on the roof surface. Generally the collectors are placed on a waterproof membrane on top of the roof sheathing. The finished roof surface, together with the necessary collector structural attachments and flashing, are then built up around the collector. A weatherproof seal between the collector and the roof must be maintained, or leakage, mildew, and rotting may occur.

Fig. 9-6. Collector mounting options.

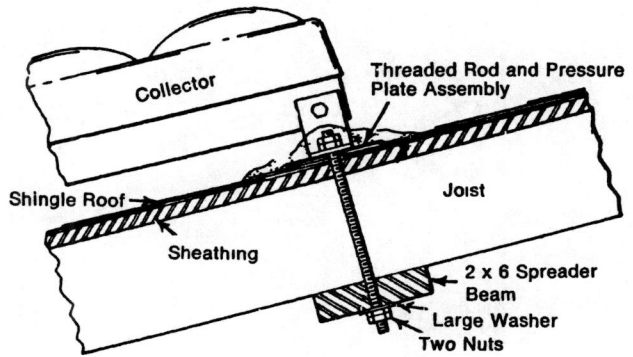

NOTE: Roof pitch is the same angle as the collector tilt.

Fig. 9-7. Stand off mounting.

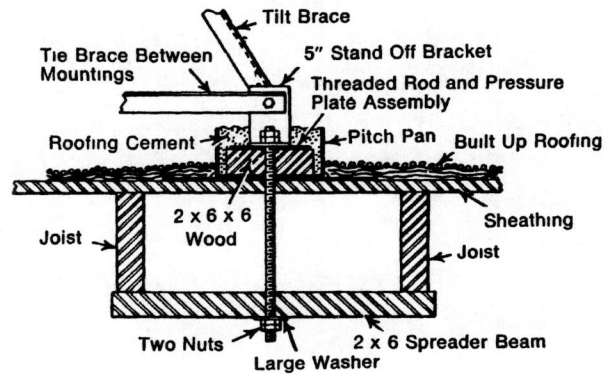

Fig. 9-8. Flat roof rack mounting. Note use of pitch pan.

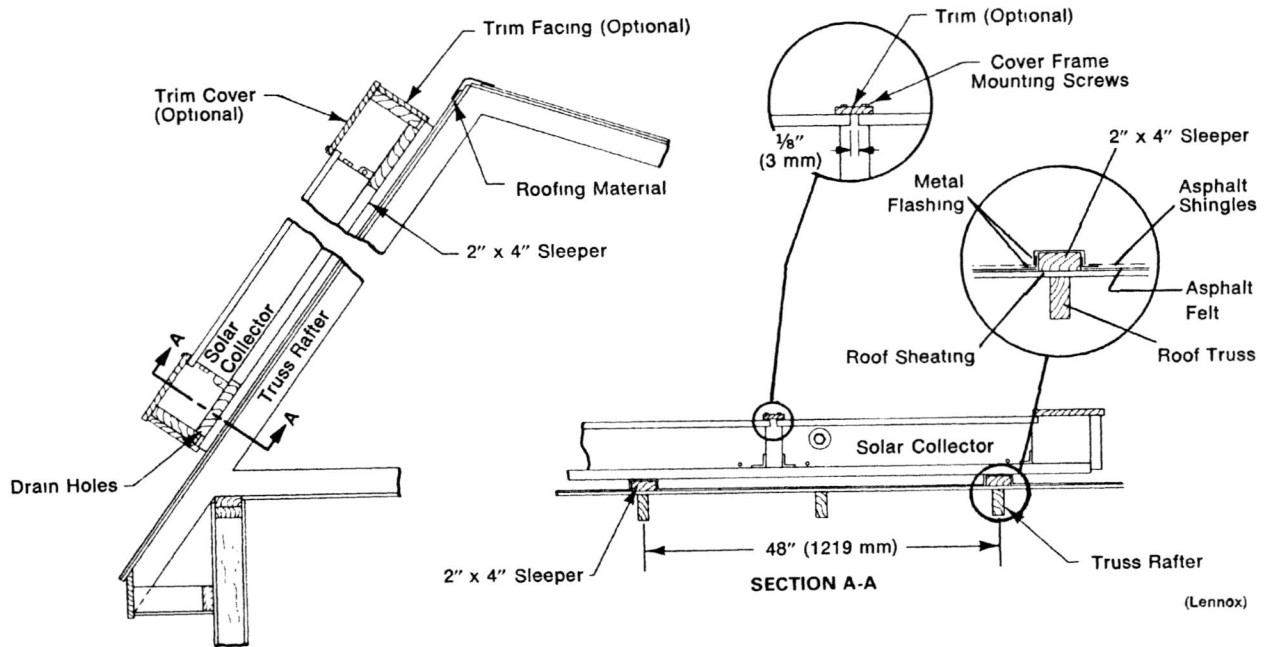

Fig. 9-9. Wood frame mounting (2 x 4 sleeper stand offs.)

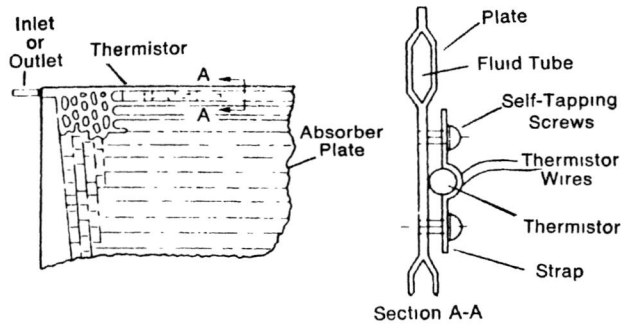

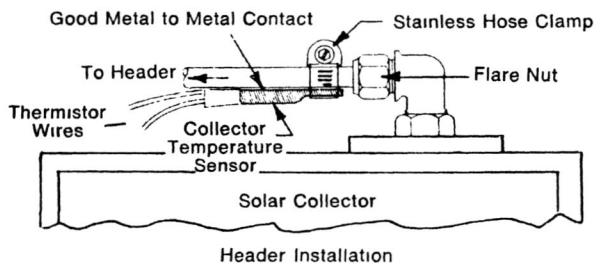

Fig. 9-10. Collector sensor installation.

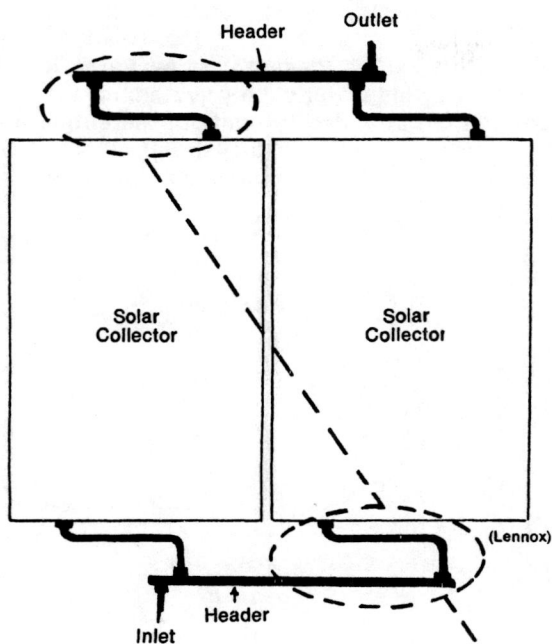

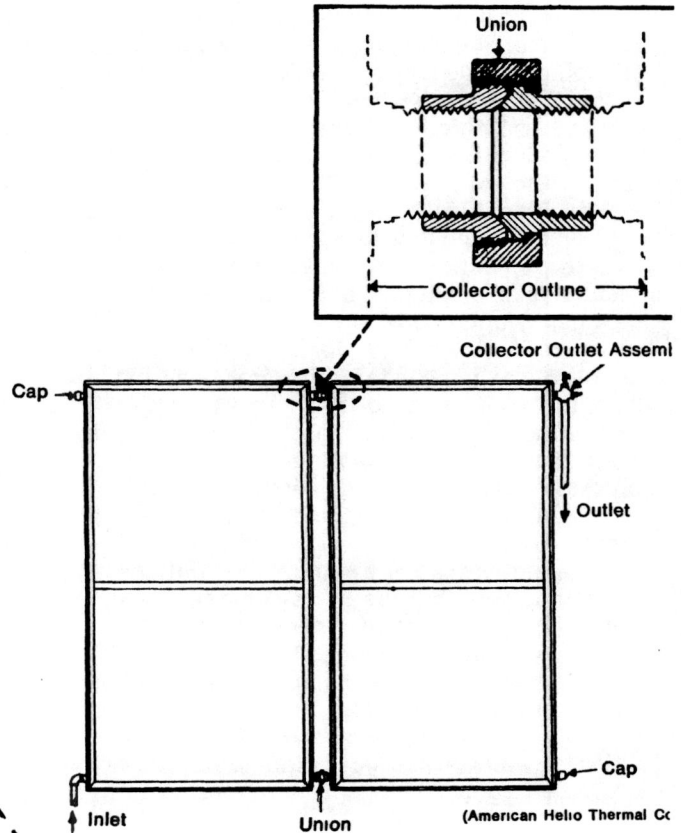

Union

Collector Outline

Collector Outlet Assem

Cap

Outlet

Cap

Inlet

Union

(American Helio Thermal Co

Header

Outlet

Solar Collector

Solar Collector

(Lennox)

Header

Inlet

ques. Some installations use high pressure rubber hose to attach the manifold to pipes or to the collectors, but hoses will deteriorate over time. From the collector inlet side (near the eave), one liquid line of the circuit is fabricated to the equipment room, and piping from the outlet manifold extends into the attic (near the ridge) where an air vent valve (Figure 9-12) is installed at the highest point in the system to remove air from the circuit. Figure 9-13 illustrates several roof penetrating practices. The outlet line then returns to the equipment room. Three possible array circuits are shown in Figure 9-14. As in ordinary hydronic design, the reverse return circuit (No. 2) is generally preferred because of balanced flow paths.

At this point, fabrication practices for the collector loop circuit in a liquid system differ between the closed- and open-loop design. The closed-loop is a system used where sub-freezing temperatures require that a heat transfer fluid or antifreeze mixed with water in the collector, circulating pump, and heat exchange circuit as shown in Figure 9-15.

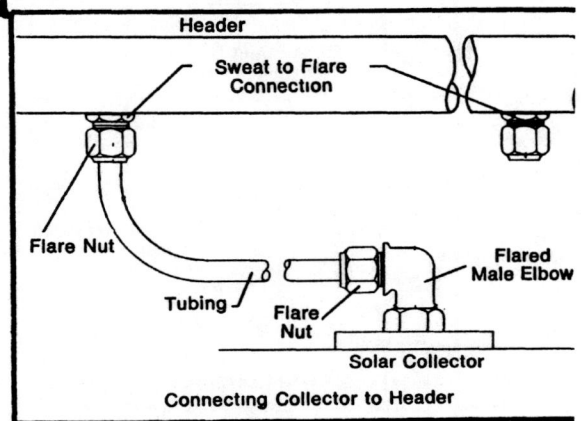

Header

Sweat to Flare Connection

Flare Nut

Tubing

Flare Nut

Flared Male Elbow

Solar Collector

Connecting Collector to Header

Fig. 9-11. Typical collector assembly procedure (eac manufacturer may design a unique system).

Centrifugal pumps should be located with at least five feet of water pressure on the intake side of the pumps. This may be difficult to accomplish in some installations where the heat storage unit is placed underground. However, it is important that this condition be met so that the pumps can operate efficiently.

Specifications for the circulating pump are based on the liquid flow-rate in the collector to optimize the collection of solar radiation. An expansion tank is also installed at this time. It may provide a handy opening where antifreeze can be added as well as be a reservoir for added liquid if needed. The expansion tank also provides a place where the fluid in the circuit can expand as it heats without increasing the pressure of the liquid in the system. The other part of the closed-loop system can include a domestic water heating sub system. The key piping and components in a typical water heating sub system are shown in Fig 9-16. Again, there are any number of variations of this arrangement, but the sub system in Fig 9-16 minimizes the amount of standby fuel input required. One of the advantages to the closed-loop system is that only a small amount of antifreeze is needed in the collector loop. This is a substantial cost reduction factor, since a thousand or more gallons of water in the heat storage unit do not have to be treated to be kept from freezing.

In areas where sub-freezing temperatures are rare, an open-loop system can be used. This type of solar heating system design uses water as it comes from the cold water supply system for the structure and is circulated throughout the entire circuit as illustrated in Figure 9-17. It is essential that piping and collectors drain completely to

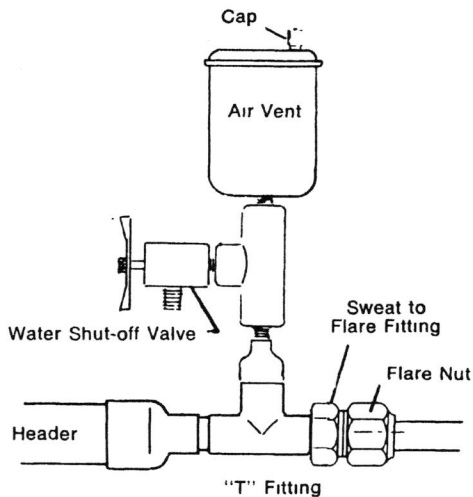

Fig. 9-13. Typical roof penetration practices.

Fig. 9-12. One type of air vent hookup. Locating vents and valves outdoors does pose the threat of freezeups and the opportunity for a valve to leak unnoticed. If outdoor locations are deemed necessary, use best quality vent and valve.

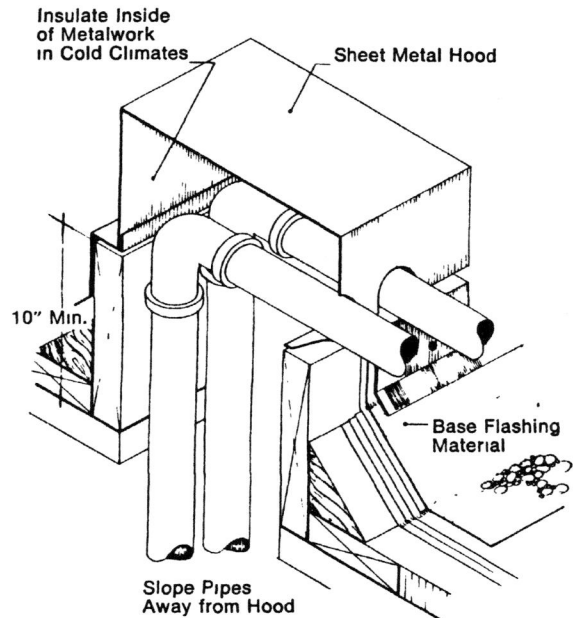

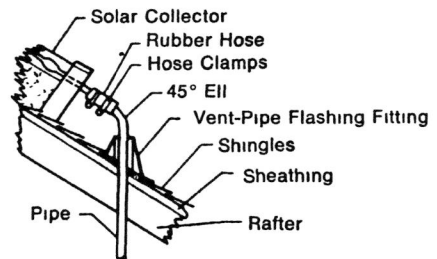

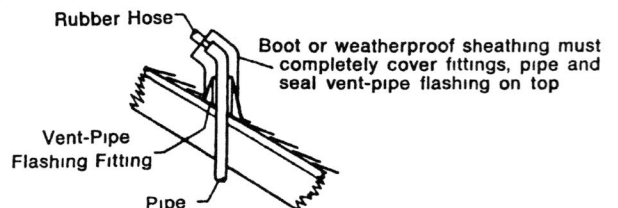

avoid freezing and boil-outs of liquid trapped in collectors.

Both systems, open or closed, make use of (1) various valves to divert the direction of liquid flow, (2) check valves, (3) pressure reducing valves, (4) flow control valves, (5) circulating pumps, (6) flow rate valves, (7) expansion tanks, (8) heat exchangers, (9) getter columns (for all-aluminum collectors), and (10) drains and other components. Most of these are located in the equipment room.

Properly installed piping is also a concern. Copper or high temperature plastic* pipe can be used to install the system. In either case, the pipe must be insulated with code-approved pipe insulation. Neoprene foam (a minimum of one-half inch thick) is an example of an acceptable insulation material (R-4 to R-7).

It is essential that all horizontal piping should be pitched upward at least ¼ inch per 10 feet of run to insure adquate drainage. The size of the pipe selected should be great enough to insure that the water velocity does not exceed five feet per second. Table 9-3 provides information about specific flow rates of common size pipes.

*Consult plastic pipe manufacturers for suitability at high temperature and pressure.

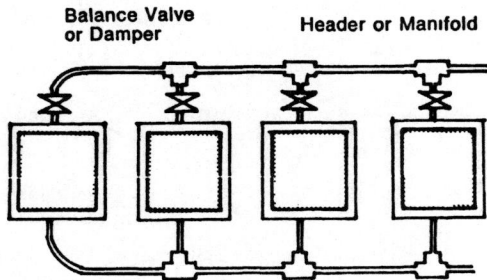

Balance Valve or Damper **Header or Manifold**

1. Parallel Flow — Direct Return — A direct return distribution circuit circulates the transfer medium from the bottom of the collector to a return header or manifold at the top. This arrangement may cause severe operating problems by allowing wide temperature variations from collector to collector due to flow imbalance. Although the pressure drops across each collector are essentially the same and at the same flow rate, high pressure drops occurring along the supply/return header or manifold will cause flow imbalance. This problem can be reduced by sizing each header for minimum pressure drop although this may be prohibitive because of economic and space limitations. Even manual balancing valves may be difficult to adjust, so automatic devices or orifices might be required for efficient system performance. Provisions must also be made to measure the pressure drop in order to adjust the flow rate to prevent collectors closer to the circulating pump from exceeding design flow rates and those farther away from receiving less.

Table 9-3. Pipe diameters for specific flow rates.

Schedule Pipe Size	Gallons per Minute	Velocity FPS	Pressure Drop per 100 Feet PSI
⅜	2	3.36	6.58
½	4	4.22	7.42
¾	8	4.81	6 60
1	15	5.57	6.36
1¼	25	5.37	4 22

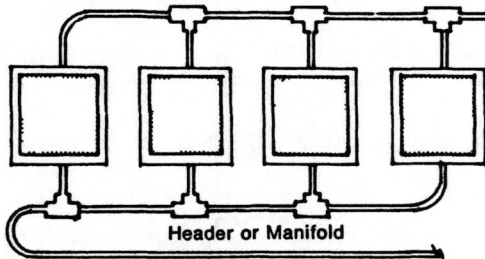

Header or Manifold

2. Parallel Flow — Reverse Return — Reverse return piping systems are coisidered preferable to direct return for their ease of balancing. Because the total length of supply piping and return piping serving each collector is the same and the pressure drop across each collector is equal, the pressure drop across each manifold are also theoretically equal. The major advantage of reverse return piping is that balancing is seldom required since flow through each collector is the same. Provisions for flow balancing may still be required in some reverse return piping systems depending on overall size of the collector array and type of collector.

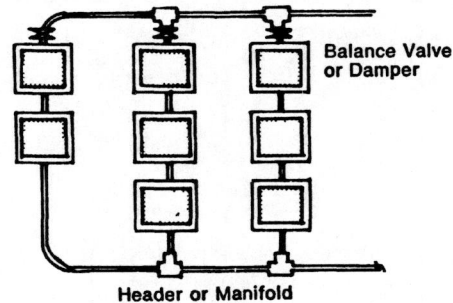

Balance Valve or Damper

Header or Manifold

3. Series Flow — Series flow is often used in large planar arrays, to reduce the amount of piping required, by allowing several collector assemblies to be served by the same supply return headers or manifolds. Series flow can also be employed to increase the output temperature of the collector system or to allow the placement of collectors on non-rectangular surfaces. Either direct or reverse return distribution circuits can be employed, but unless each collector branch has the same number of collectors, the reverse return system has no advantage over direct return — each would require flow balancing.

Fig. 9-14. Collector piping circuits.

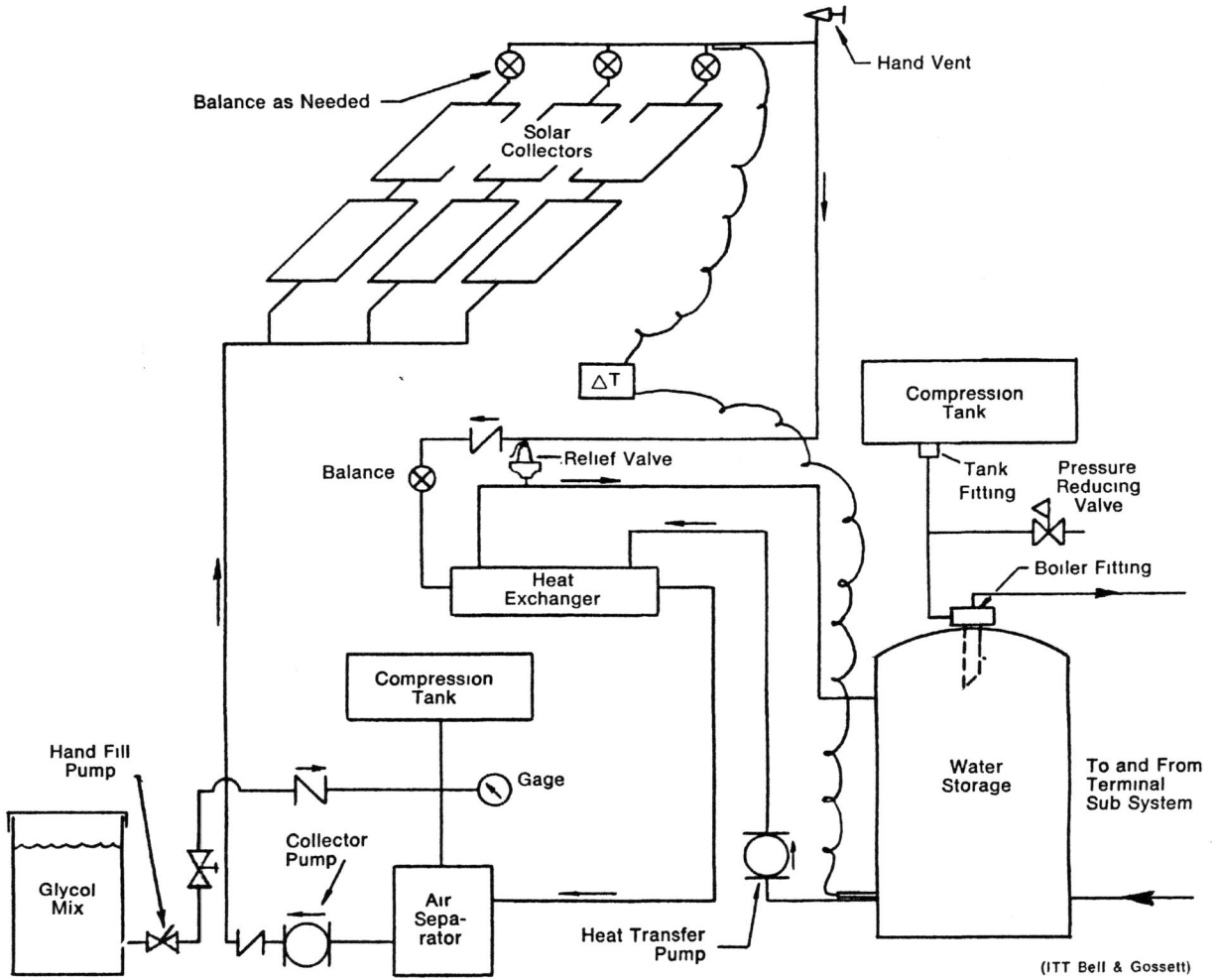

Fig. 9-15. Piping arrangement for a closed loop system.

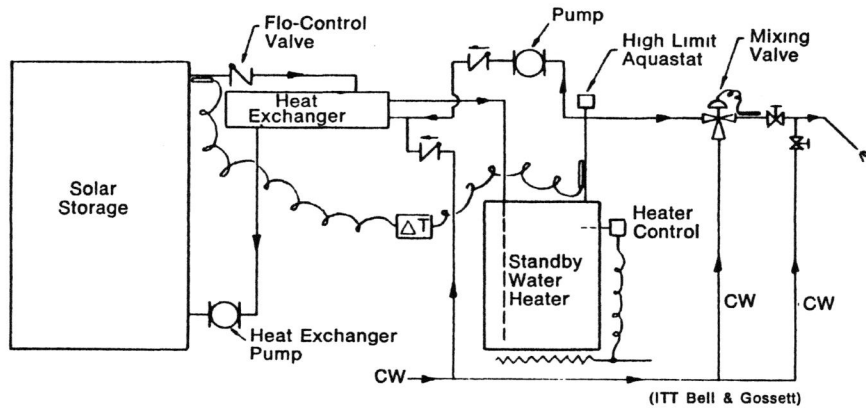

Fig. 9-16. Domestic water heating sub system.

To insure a high quality piping installation, the following is presented:

Ten Commandments for Good Piping Practices

1. The ends of all pipe or tubing should be reamed.

2. Keep branches as short and as uniform as possible.

3. Use a minimum number of fittings, and use eccentric reducers to join different pipe sizes.

4. The use of soft copper tubing eleminates the need for elbows.

5. Use a bending tool for making all bends in the tubing.

6. Allow for expansion in all long runs of pipe or tubing. Provide ample clearance around all pipe or tubing which passes through floors and walls.

7. Fasten the tubing to the joists with either staples or straps, leaving sufficient clearance between the tubing and joists to permit expansion.

8. Install circuits and mains horizontally and as direct as possible, keeping as close to the wall as is practical.

9. When more than one circuit is installed, place square head cocks in each circuit for balancing.

10. Keep the inside of pipes and tubes clean.

Pipe joints are extremely important to the operating efficiency of the liquid system. When using galvanized piping, the threads should be coated with pipe joint compound to insure that they will not leak when tightened. For copper piping, it is necessary to clean both the pipe and fitting as shown in Figure 9-18 with some abrasive paper before making the soldered connection.

Several manufacturers insist that *no* hand shut-off valves be placed in the collector loop circuit. This would eliminate the chance that a homeowner (or inexperienced installer) may inadvertently restrict the flow of liquid and perhaps damage the collectors as a result of excessive temperatures. Still others insist that temperature probes or sensors be installed in a Tee fitting rather than protrude into the collector loop circuit and impede fluid flow.

Expansion tanks must be adequately sized. Experience indicates that larger than normal expansion tanks may be required in solar systems relative to conventional water filled hydronic heating systems. Most expansion tank manufacturers provide simple tank sizing information based on volume of water or special fluid in the system and temperature rise expansion of the liquid.

The installer must be alert to the problems of an electrolytic action that causes rapid decomposition of dissimilar metals especially copper and aluminum. (See SMACNA Section 18.) For example, a dielectric union could be installed between iron and copper components. Short sections of rubber tubing that will withstand pressures of over 150 psi can also be used to minimize electrolysis. Electrolytic action will cause erosion of the metal and the build-up of a calcium carbonate deposit near the joint that will restrict liquid flow rates. For all aluminum collectors, installation of a getter column (Figure 9-19) in the collector loop will also help to minimize the problem of deterioration of some components.

Sensors manufactured with pipe threads must be made watertight when they are installed in tanks and pipe fittings. Bulb wells, where used, must also be watertight.

Once the rough-in plumbing has been completed, the system must be systematically filled and tested for leaks. Collectors, tanks, pipes, and pumps, are filled with water and pressurized to the limits that the component manufacturers specify. When testing has been completed, the parts of the system that must be protected from heat loss are ready to be insulated. Note: this activity should *not* be performed in cold weather if there is no heat in the building. Also, it may be necessary to drain the system after testing if there is a possibility of subfreezing weather before the auxiliary heating system becomes operational. An alternate procedure would be to use *air* to pressure test the system.

Electrical Service. The next phase of construction relative to the solar heating system is electrical service. The first step is to rough-in the service wiring to the location of the main control panel. The second step is wiring for the sensors from their various locations to the control panel.

After the rough-in electrical service is installed, the solar system control panel can be assembled. (See SMACNA Sections 16.1-16.8) This panel consists of (1) differential thermostats, (2) controls, (3) relays, and (4) other electrical and electronic equipment prescribed by the heating system manufacturers. Be sure that the components on the panel are well ventilated because the heat generated internally must be dissipated through heat sinks.

All sensors should be wired to their respective control terminals at this time.

Next, the structure must be weatherized. Exterior siding is applied and caulked. Solar heating pipes, tanks, and other components are insulated and inspected. Then the remainder of the structure—walls, ceilings, heat ducts—should be insulated. The amount of insulating material used is determined by the geographical area of the building site and cost factors.

The auxiliary furnace should be installed next. This is important so that there will be heat in the structure when dry wall or plaster is applied if construction is underway during cold weather.

The final phase of construction for the liquid circuit solar heating system is to complete the installation of the domestic hot water system. This involves connecting and leak-testing the preheat and auxiliary water tanks to the cold and hot water supply pipes and the gas, oil, or electrical utility used as an auxiliary water heating energy resource.

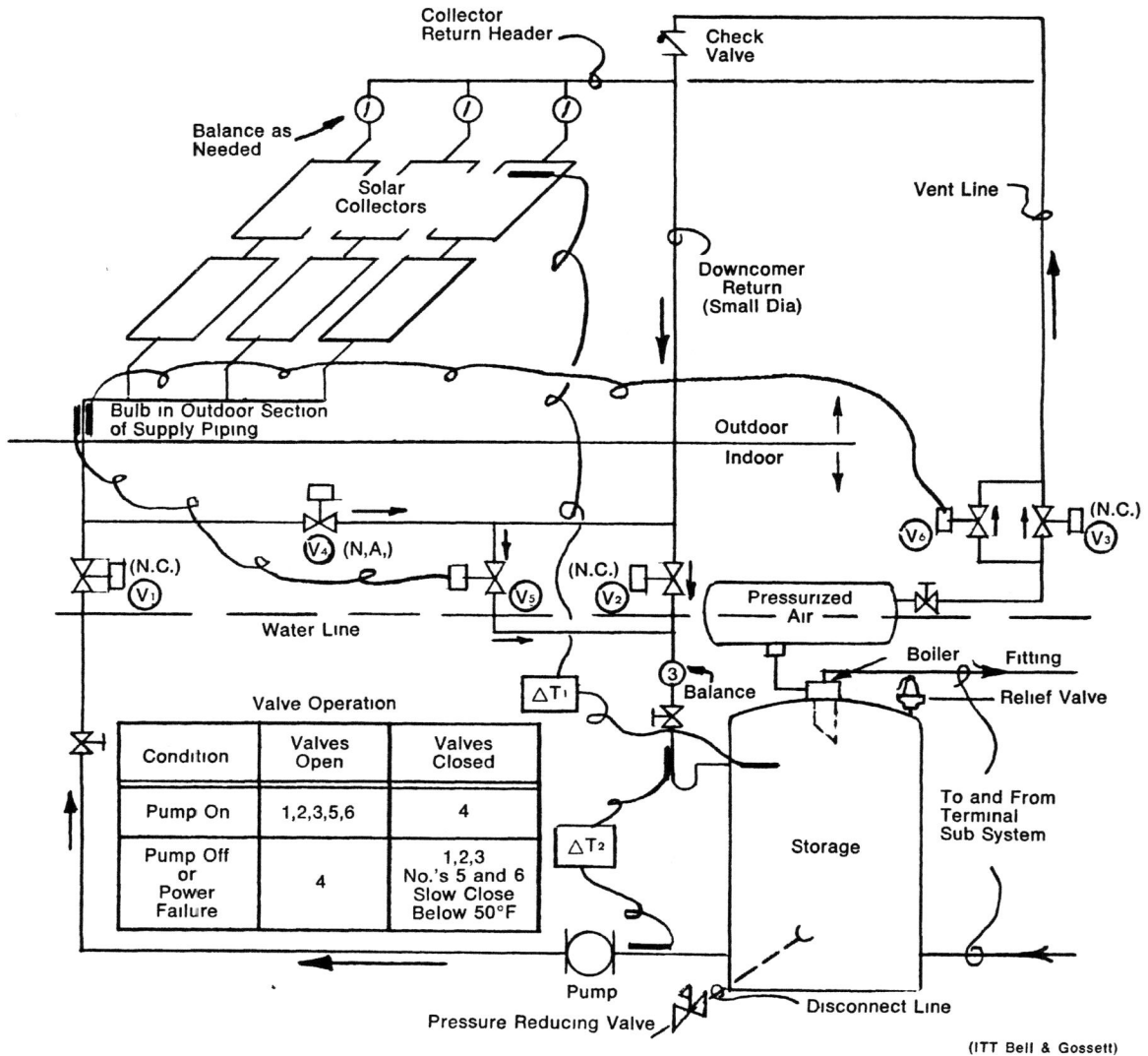

Valve Operation		
Condition	Valves Open	Valves Closed
Pump On	1,2,3,5,6	4
Pump Off or Power Failure	4	1,2,3 No.'s 5 and 6 Slow Close Below 50°F

(ITT Bell & Gossett)

Fig. 9-17. Drain down on open loop system details.

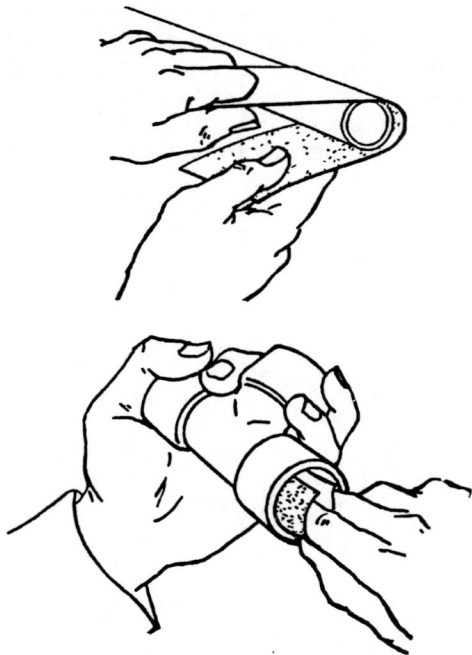

Fig. 9-18. Soldered joint preparation.

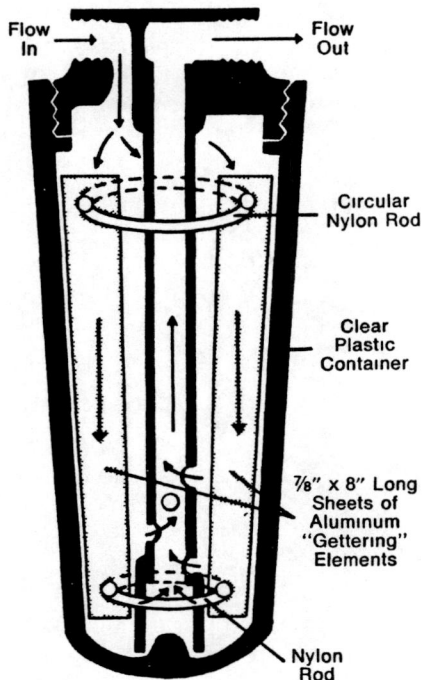

Fig. 9-19. Ion "getter" column.

Air Systems

Solar-assisted air heating systems are essentially the same in operation as the liquid system. Only the components are changed. Collector designs are different. Pipes become ducts. Fans replace pumps and dampers replace valves (See Figure 9-20).

Scheduling construction for the air system is comparable to that for the liquid system. The installer needs to know what the physical specifications are for each component and where the builder plans to have the device located.

Heat Storage Unit. (SMACNA Section 11.5, 11.6 and 11.7) Rock filled heat storage units for air circulating solar heating systems are normally constructed on-site. They can be made of poured concrete, concrete blocks, reinforced wood framing, or prefabricated steel boxes. In many instances, one or more of the sides of the concrete storage "box" may be an integral part of the foundation of the structure.

Footings for the heat storage unit are poured simultaneously with the basement footings. Keep in mind that rock has one-fifth of the heat storage capacity as water and that rock is almost twice as heavy as water. Therefore, the unit with a given storage capacity will be about five times as large, by volume, and must support almost twice as much weight as a liquid solar heating system.

The heat storage container can be fabricated of concrete blocks with 3/8 inch reinforcing rods,

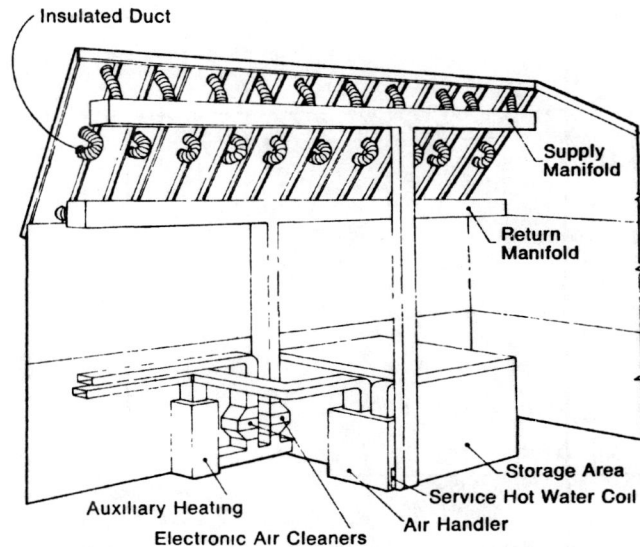

(Research Products)

Fig. 9-20. All air system.

9-15

poured-in-place concrete, or wood framing in an out-of-the way location. The unit may be designed horizontally (see Figure 9-21) or cubical (Figure 9-22).

If wood frame construction is selected, it should be constructed using a minimum of 2 x 4's on 16 inch centers with ½ inch plywood on both sides of the studding. Full insulation (e.g., fiberglass or rockwool) should be placed between the studs. The inside of the plywood should be covered with fire-rated sheetrock. Inlet and outlet duct openings must be provided for when the framing is being constructed. Also, filters should be installed on either side of the storage bin to reduce dust accumulation.

The builder must keep in mind that an access door must be provided in the top to allow for inspection of the rocks. This is essential because, when rocks become coated with dust, they lose some of their heat absorbing capabilities.

The box can then be filled with washed rock (up to fist size). Filling the box should be done in such a way that dust is minimized. The system may be designed for a storage unit heat sensor to be placed in the center of the unit rather than the top or bottom. In this case, filling will have to be stopped while the sensor and necessary conduit and wire are put in place. (The sensor is placed near the bottom in most cases.) Then the filling

can continue. Because of the weight of rock, concrete heat storage units should be limited to about 6 feet in height with about 5 feet of rock filling. The insulated and sealed top, not necessarily made of concrete, should be air tight and put in place as soon as possible to prevent rocks from getting dusty.

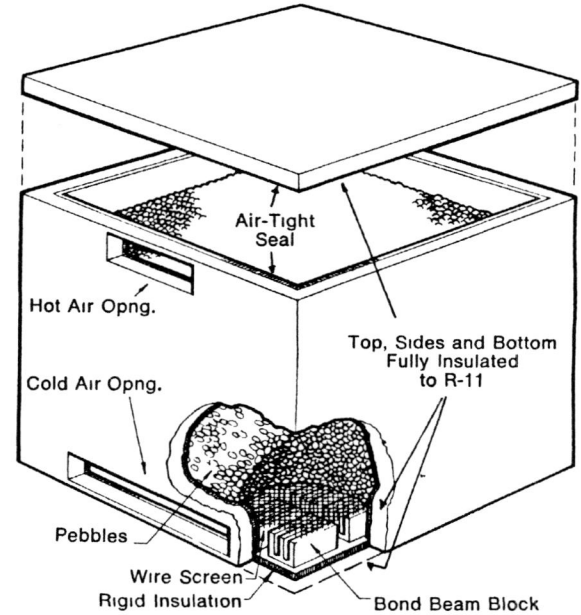

Fig. 9-22. Cubical storage.

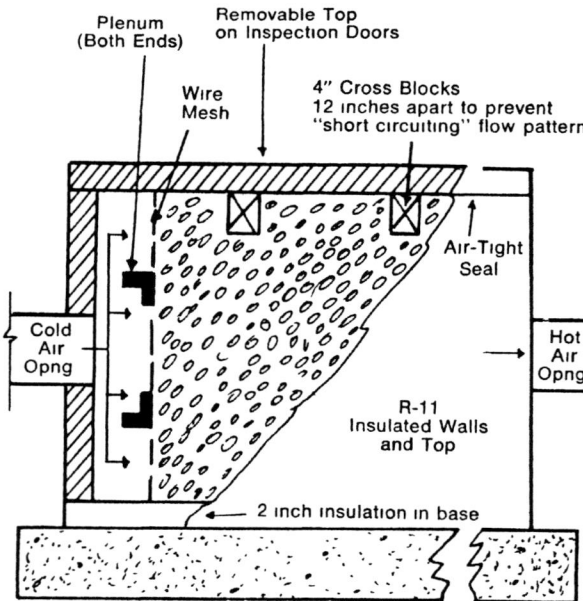

Fig. 9-21. Horizontal flow pebble-bed.

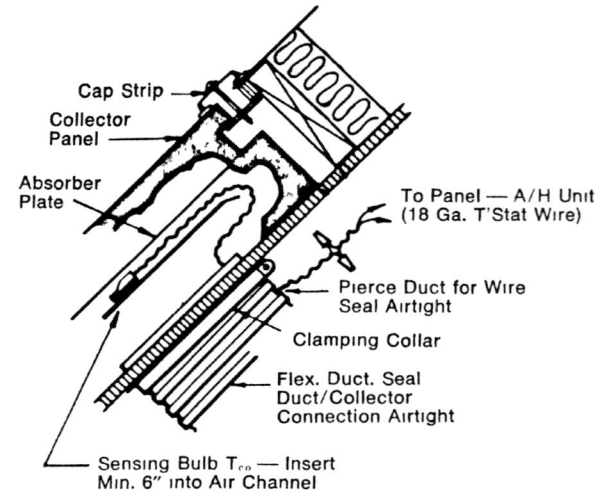

(Solaron)

Fig. 9-23. Sensor location in air collector.

Preparations for Collector Installation. Construction scheduling continues from this point with the placement of girders, floor joints, and subflooring. Exterior walls are then framed, sheathed, and erected. Following these activities, the collector supports, roof rafters or trusses, sheathing and felt are set in place. Interior partitions that are typically non-load-bearing may need to be fabricated if added support is required for the collectors.

After the collectors are inspected, the absorber plate sensor is installed in the inlet or outlet collector (Figure 9-23). Then the collectors are ready to be hoisted into place. The array of collectors will be mounted on the roof as predetermined during the design phase of the project. A chalk outline of the actual array should be made on the roofing felt as well as sheathing to determine location of cuts for duct connections.

Collector Array Installation. The collectors are mounted one at a time on the roof. (See Figure 9-24). A typical layout is shown in Figure 9-25. The holes in the sheathing are placed so they will line up with the inlet and outlet ducts for each array. Note that the openings for duct manifold-to-collector connections are typically on the back of the collector casing. Panel-to-panel series connections are made directly to each other by means of flanged openings with a mating gasket. Unused ports are capped to prevent air leakage.

As the collector array is put in place, the flashing and roofing can be installed. If the collectors are an integral part of the roof, the flashing will cover the sides and top edges of the collector panels. This procedure will prevent water from entering the duct holes in the roof. Narrow cap strips of flashing material keep water from getting between the collectors. In addition, the cap strips hold the glass cover plate in place (see Figure 9-29). When the entire array has been sealed with flashing and cap strips, work on the collectors can continue with the installation of ducts leading to the equipment room.

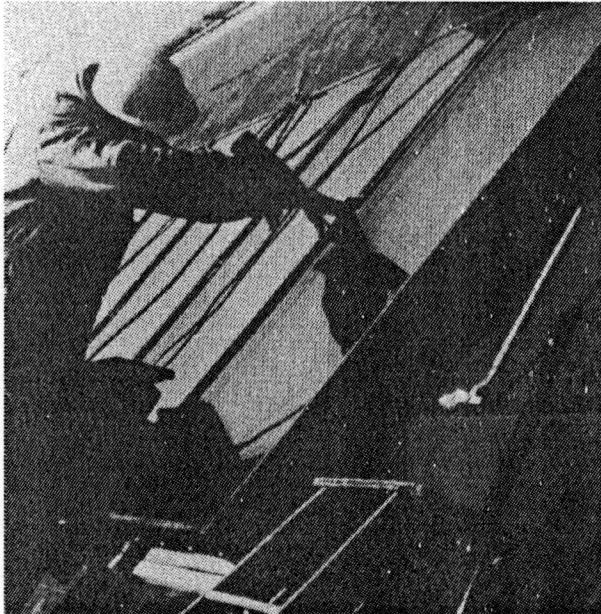

(Solaron)

Fig. 9-24. Collectors being fastened to roof. Note "pull up" tool used to draw collectors together for proper spacing and to seal interconnecting ports.

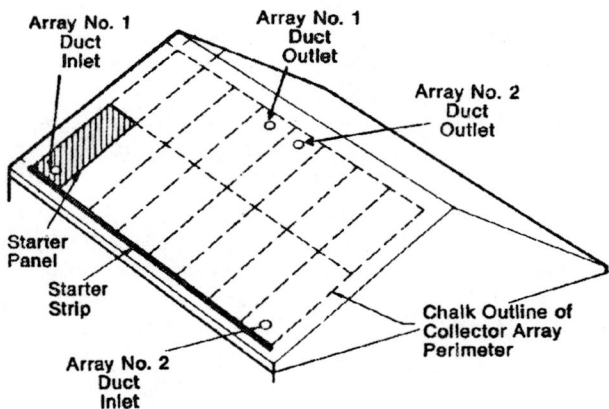

Fig. 9-25. Roof outline of collector array.

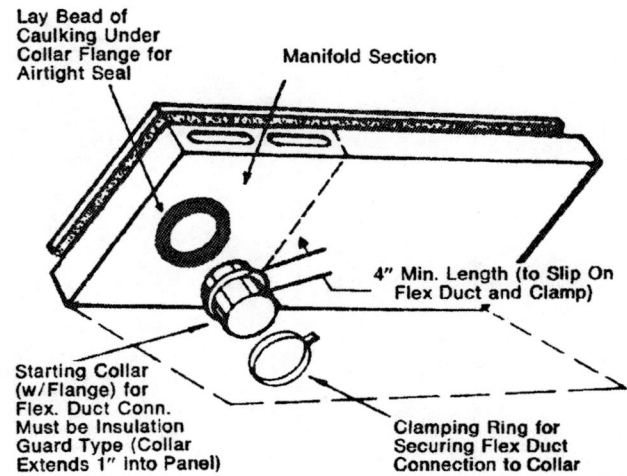

(Solaron)

Fig. 9-26. Collector duct connection.

9-17

Duct Work Installation. This phase of the construction schedule is extremely important. Its function for the air system compares to the plumbing activities in the liquid system. (Refer to Figure 9-28.) All duct work should be installed according to SMACNA Section 7.1-7.13 standards.

Rough-in duct work activities involve all ducts from the collector array to heat storage, to auxiliary heat, and to air distribution systems. The process can start in the attic at the collector inlet and outlet ducts. The collectors may be connected to each other in a number of flow circuit

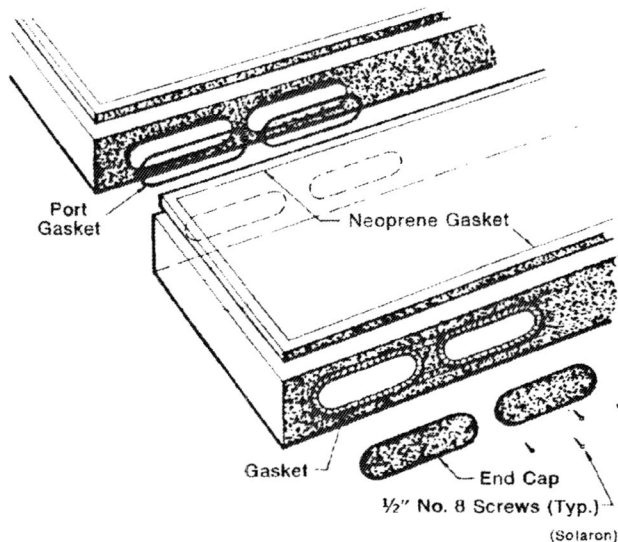

Fig. 9-27. Collector "port" assembly.

Port Gasket

Neoprene Gasket

Gasket

End Cap

½" No. 8 Screws (Typ.)

(Solaron)

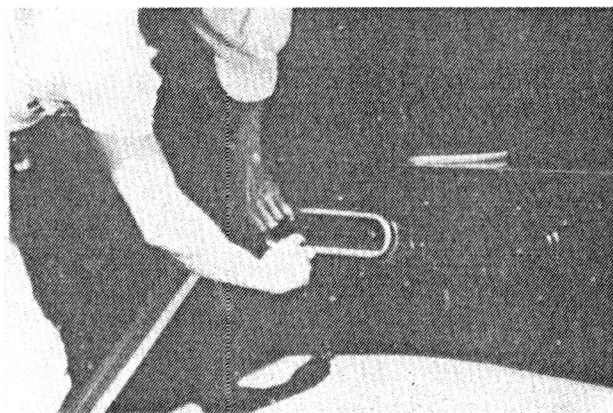

(Solaron)

Fig. 9-28. Gasket is centered on collector port before hoisting and positioning panel alongside a companion module.

patterns. See Figure 9-30. Collector manufacturers specify the number of manifold connections (inlets and outlets) required for each array pattern. There may be more than one array on a roof. When this is the case, there is a need for separate inlet and outlet manifolds. Manually operated dampers may have to be installed to insure even flow-rate through each array. Remember that proper flow is critical to collector efficiency.

Ducts leading from the (1) collector inlet to the heat storage unit and (2) collector outlet to the equipment room are installed next. Space for these ducts as they extend from the roof to the equipment room and the storage unit (probably in the basement) will require one or more square feet of floor space. The building plans should show the size and location where the ducts will pass through the various ceilings and floors. Ducts may be made of fiberglass ductboard or of sheet metal lined on the inside or covered on the outside with fiberglass insulation. Fiberglass ductboard should be used in places where it is will protected because it can be damaged easily. Fiberglass duct joints must be joined and sealed with the recommended adhesives. Metal ducts may be jointed with drive clips (Figure 9-32) and covered with joint tape so they will be air-tight. Bends or elbows in the sheet metal ducts should contain turning vanes to minimize resistance. Ducts can be supported by using standard mounting procedures.

Connections between ductwork and blowers should be made with flexible connectors. There are two reasons for this practice. One is that any noise from the blower will be dampened by this fabric-type connection. (SMACNA Sections 15.3 and 7.8) The other reason is that the blower can be removed easier for servicing because these flexible fabric-type connectors are held in place with screws.

The collector inlet duct extends from the roof to the heat storage unit which is usually in the basement, in a crawl space, or possibly under a garage floor. The collector outlet extends from the roof to the equipment room.

Equipment Room Systems and Controls. The space used within a structure to house the necessary equipment for an air system may be extensive. This is particularly true if the heat storage unit is placed there as shown in Figure 9-33.

In the equipment room, the duct from the collector is attached to the air handling module. See Figure 9-35. This unit functions in a manner similar to the electrically controlled valves in the liquid system. The module automatically direct air

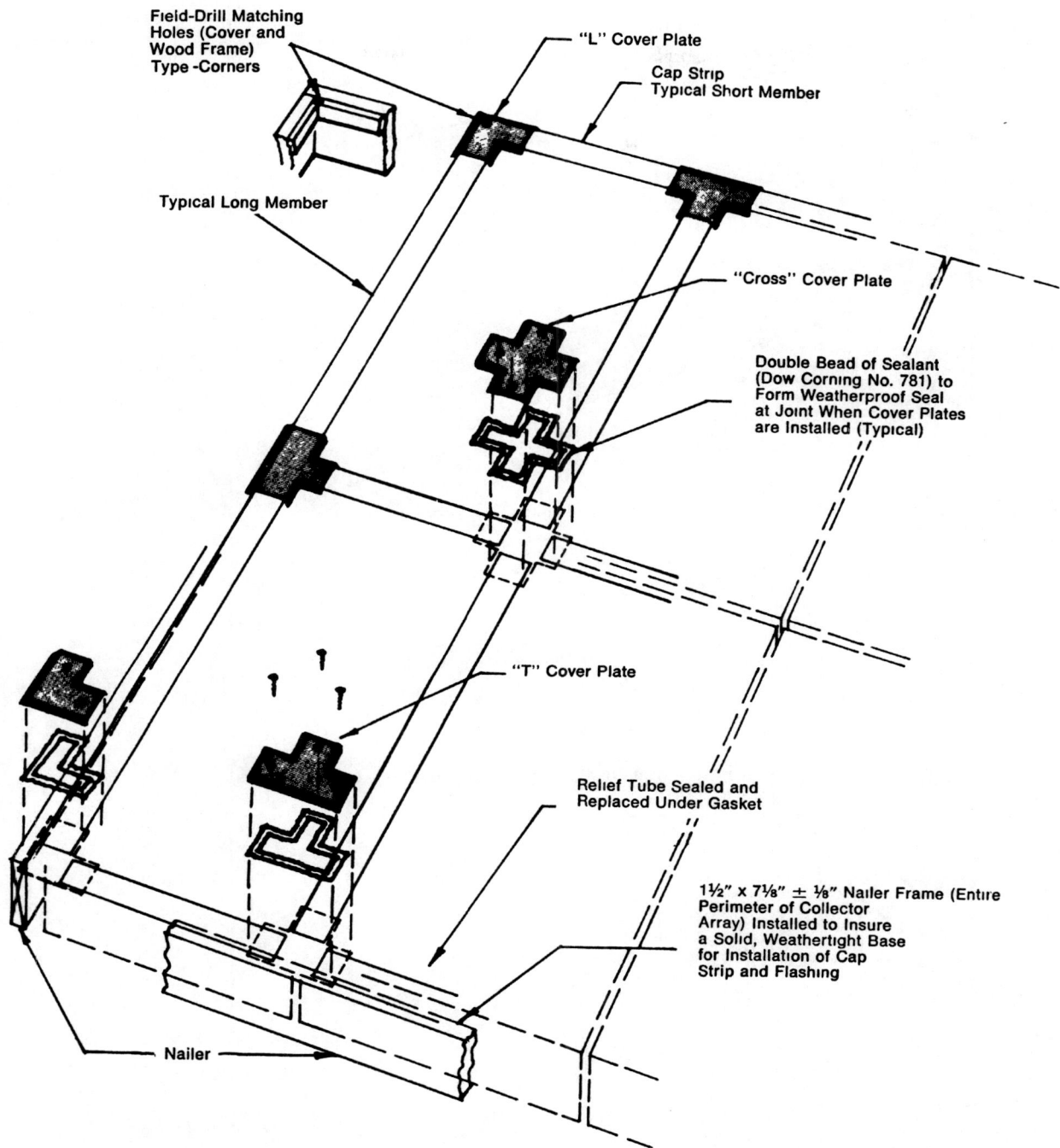

Field-Drill Matching Holes (Cover and Wood Frame) Type -Corners

"L" Cover Plate

Cap Strip Typical Short Member

Typical Long Member

"Cross" Cover Plate

Double Bead of Sealant (Dow Corning No. 781) to Form Weatherproof Seal at Joint When Cover Plates are Installed (Typical)

"T" Cover Plate

Relief Tube Sealed and Replaced Under Gasket

$1\frac{1}{2}$" x $7\frac{1}{8}$" $\pm$ $\frac{1}{8}$" Nailer Frame (Entire Perimeter of Collector Array) Installed to Insure a Solid, Weathertight Base for Installation of Cap Strip and Flashing

Nailer

Fig. 9-29. Cap strip installation.

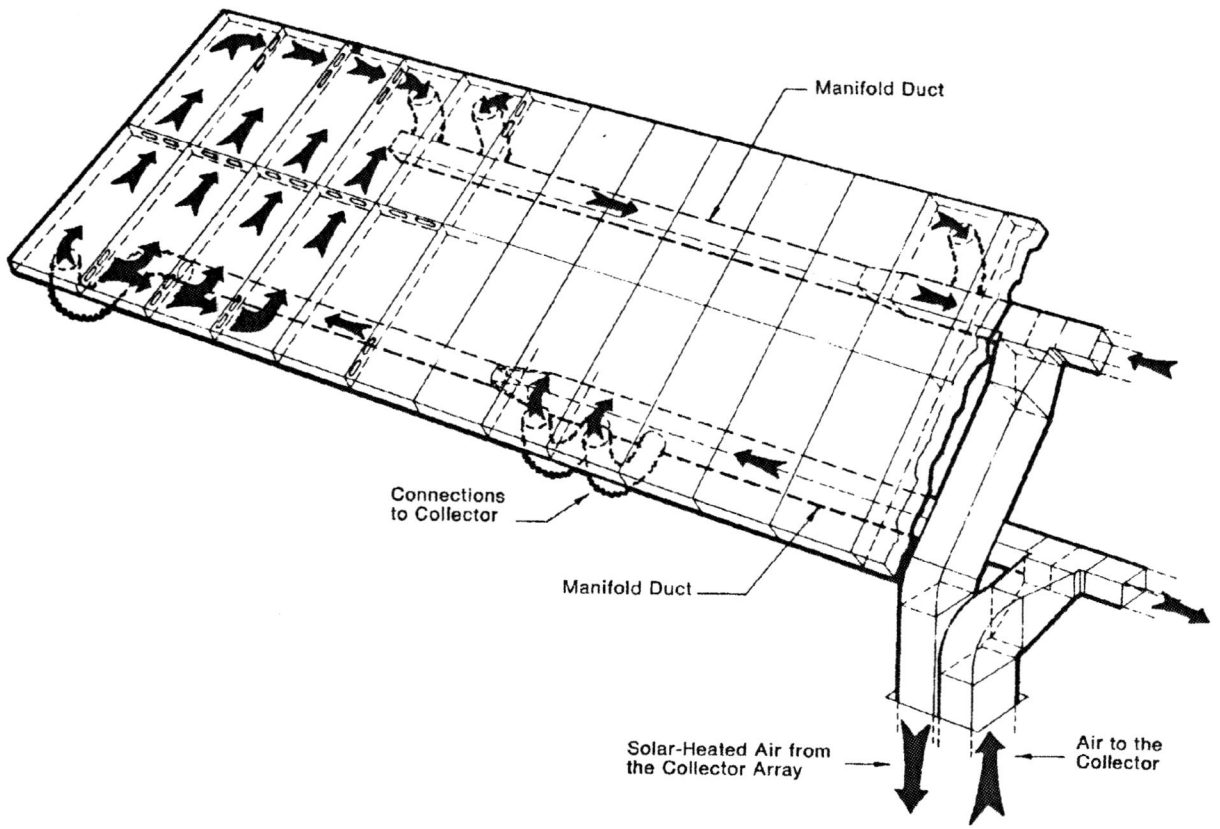

Typical Collector Installation

Fig. 9-30. Collector array details.

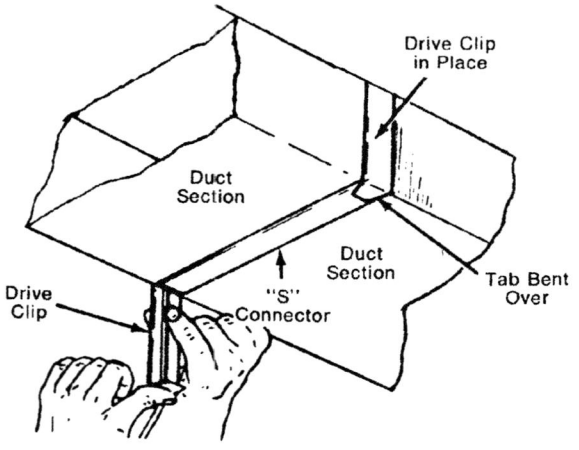

Fig. 9-32. Slip and drive duct connections should be taped to reduce leakage.

(Solaron)

Fig. 9-31. Exterior view of collector array.

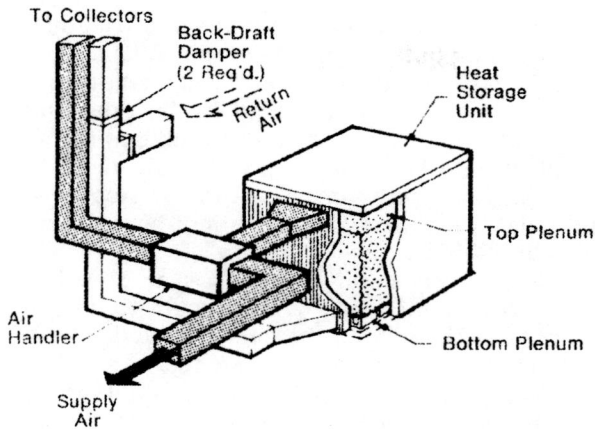

Fig. 9-33. Equipment installed indoors.

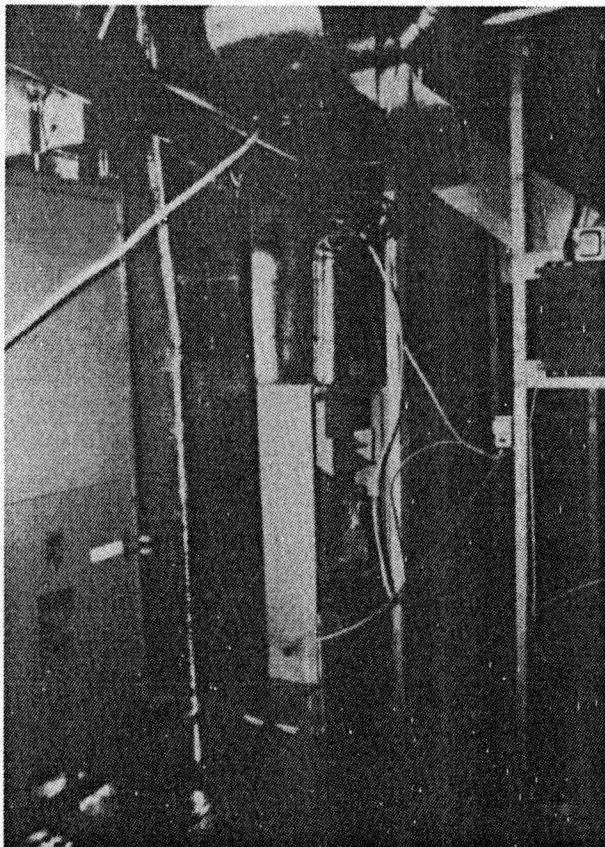

flow throughout the following operational modes: (1) space heating from the collectors, (2) space heating from the pebble-bed heat storage unit, (3) space heating from the auxiliary heating unit, and (4) heating the rocks in the heat storage unit. Internal components of this module include: (1) a blower operated by a motor with the proper horsepower rating to control a specified air flow rate, (2) four motorized dampers, (3) a 24-volt controller to which heat sensors are connected to regulate various operational modes, and (4) an *optional* pre-heat domestic hot water heating coil which may be attached to the hot air inlet on the module.

The unit is designed to operate in a horizontal or vertical position. See Figure 9-36.

A filter is essential to the efficient operation of an air solar heating system. It prevents dust, picked up in the air duct system, from coating the collector surface and reducing collector output.

(Solaron)

Fig. 9-35. Exploded view of air handler and damper system used to divert heated air from collectors to storage or to the space to be heated.

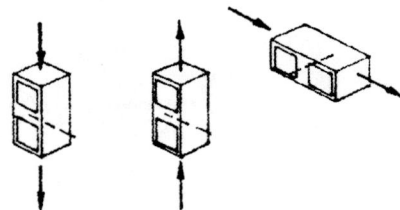

Fig. 9-34. Air handler (left) and auxiliary gas furnace (right) are shown before drywall was installed to close in equipment room.

Fig. 9-36. Most air handlers can be mounted in a variety of positions.

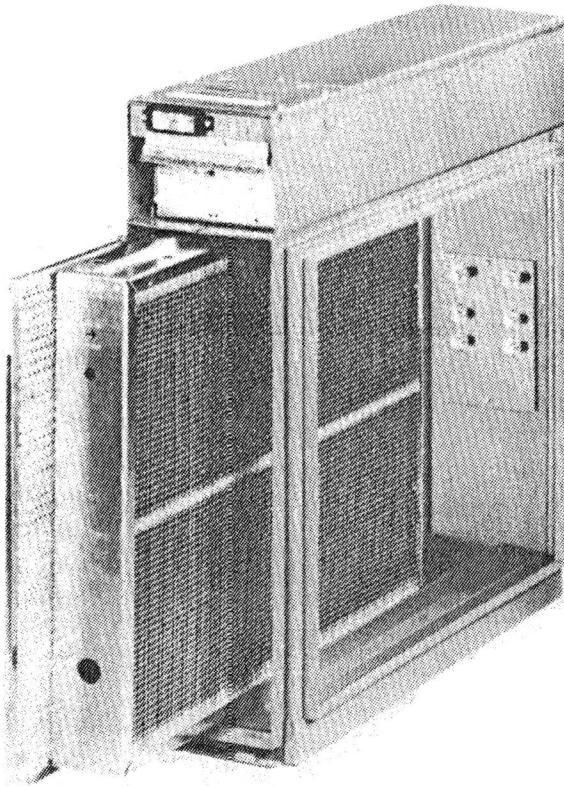

The filter should be installed in the return air duct supplying air to the inlet side of the collector and the heat storage bin. If an electronic air cleaner is desired, it should be installed in the return air duct mentioned above. Do *not* install it on the inlet of the auxiliary furnace, as the air temperatures at this location may exceed the electronic air cleaner's maximum operating temperature (usually 125°F).

A pair of back draft dampers are also needed. They prevent air from being circulated in the wrong direction during the heat-from-collector or heat-from-storage operational modes.

Most of the necessary electrical controls for the air solar heating system are in or on the air handling module. Electrical service requirements are determined by the voltage and amperage needs of blower motors.

Conventional electronic air cleaners can be used in solar systems provided they are installed in a duct where air temperatures do not exceed the operating limits of the air cleaner.

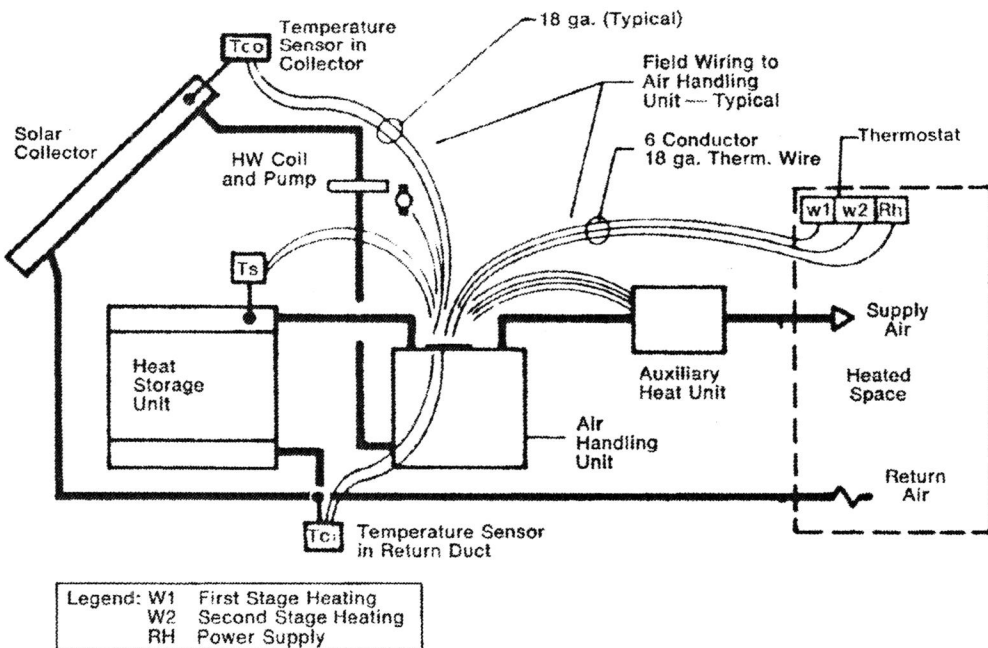

Legend: W1   First Stage Heating
W2   Second Stage Heating
RH   Power Supply

Fig. 9-37. Example of low voltage field wiring connections required for an air system.

Electrical Service. Electricity for the air heating system is used to operate blowers, motorized dampers, and various controls. Normal 120 or 240 volt electrical service is needed at the control panel for the solar heating system. The size and number of the wires installed between the control panel and the disconnect system for the structure are determined by amperage and voltages requirements of the electrical controls and circulating fans (120 or 240 VAC single phase). For additional information, see SMACNA Sections 16.1—16.8.

The control panel should be mounted in a convenient location that allows easy access for electrical switch operation. Generally, the mechanical room is the best location. Electrical service consisting of one 120 VAC circuit is ample to power the 100 VA, 120 VAC/24 VAC transformer blower unit and the auxiliary heating unit (refer to local and national building codes).

Low voltage wiring is needed to connect the room thermostat to the control panel as well as between the auxiliary heating unit, damper motors, and the control panel. Damper motors are typically low voltage.

Figure 9-37 illustrates the various low voltage field wired circuits for the system.

Figure 9-38 shows the various control panel connectors for the various controls. This particular panel may be located on the air handling unit.

Plumbing. Very little plumbing is associated with the air system. The only connections are for the hot water heating system where it is attached to the heat exchanger.

Plumbing Leak-Testing. A leak-test for the plumbing system is performed by filling the preheat and domestic hot water tanks and checking for leaks. The system must be drained after the test if freezing temperatures are expected before the auxiliary heating system is installed. An alternative to draining the system would be to delay the leak-test of the plumbing until there is heat in the structure.

Air System Leak-Testing. Testing for leaks in the air system is more difficult than in a liquid system. Water dripping from an improperly soldered joint or loose threads is easy to find. Air seepage from a duct joint is more difficult to locate.

Leak-testing will have to be accomplished by blowing air through the ducts. Leakage may be difficult to locate as the tester's hands pass over joints and seams. Another practice could be to introduce a non-toxic scented gaseous material

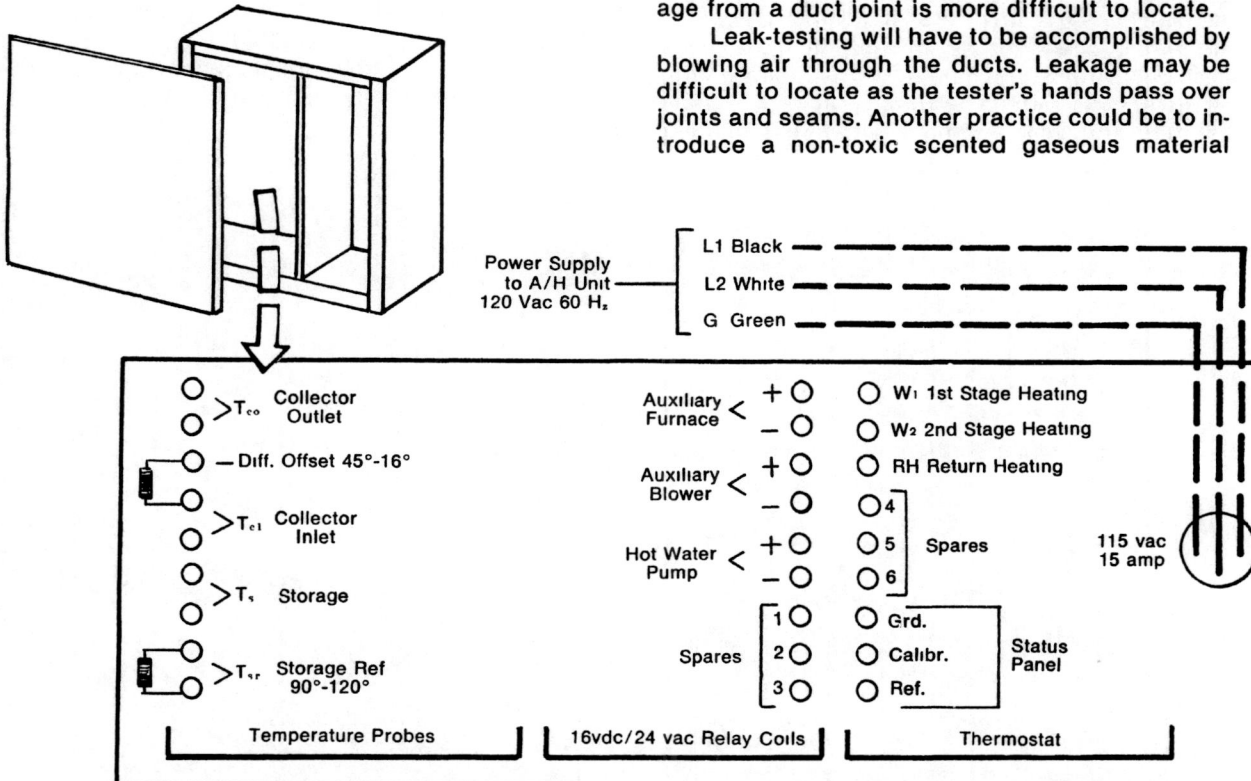

Fig. 9-38. Inside a typical control panel.

9-23

(room deodorizer) and use the sense of smell, as well as touch, to locate air leaks.

When all leaks have been sealed, uninsulated components of the entire air duct system can be insulated according to specifications. Once this has been completed, the construction schedule for the structure can proceed. Drywall can be hung. The interior and exterior can be finished. After final inspection, the structure is ready for occupancy.

Auxiliary Heating System

The auxiliary heating system used in conjunction with a solar heating system is of the same type that is traditionally installed in homes. It is connected to the solar heat system with pipes or ducts. Electrical and plumbing services to these systems would be standardized and the proper installation procedures should be followed.

Humidifiers may also be a desired element in the system. Horizontally mounted humidifiers are recommended. Locating the humidifier in a horizontal supply duct coming off of the auxiliary heating unit is ideal. Utilizing a sail-switch activated duct humidistat will simplify the wiring requirements of most installations.

Some single-stage thermostats will have to be replaced with two-stage models, although they would be located very much like they have been in the past. The multi-element thermostat should be located on an interior wall free from cold and warm drafts. Be sure adequate room air movement is present so the thermostat will provide a comfortable building temperature.

Do not locate the thermostat near lamps, heat outlets, stoves, fireplaces, refrigerators, television sets, etc. The heat given off by these appliances will not allow the thermostat to properly control the building temperature.

The duct work is also standardized to some extent. Installation of heat ducts and air returns is not part of this course but the procedures to follow are reviewed in Appendix C.

Domestic Water Heating

Guidelines for installing domestic water heating systems parallel space heating systems. Because of the interfacing of the solar system with potable water, there are frequently strict local codes that may require *double safety* measures to assure the protection of the occupants from any health hazard.

Recognizing that local codes can and do vary, some of the typical installation details for

Fig. 9-39. Conventional heating units such as this electric furnace can be used as auxiliary heat.

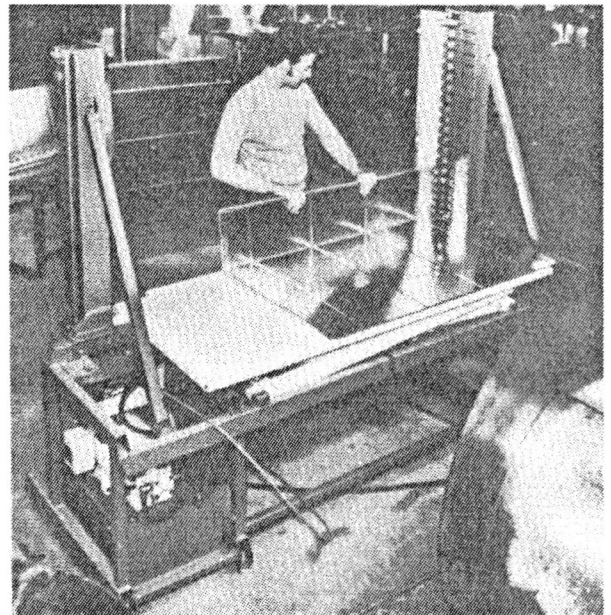

Fig. 9-40. Conventional duct fabrication techniques are used but seams and joints made air tight.

standard domestic water heating systems are presented in this section.

Closed-Loop. (SMACNA Sections 19.1 through 19.6) The following recommendations are typical of the system (See Figure 9-41):

Collectors. Mount at determined angle and pitch, as closed to tank as possible. If lines cannot pitch to drain properly, install drains and vents to alleviate air. Be sure that piping is run in a reverse/return mode. Collectors are rated 50 psi maximum.

Tank. Support tank off floor using concrete blocks if possible to allow for draining off and cor-rosion protection if basement flooding should occur. The tank should be near the existing hot water heater since the feed (supply line) to this tank is interrupting the supply line to the existing system. This will allow the solar system to act in a "preheat" mode as well as supply 100% capacity. Using a conventional water heater without a fuel system is a practical solution.

Pump. A cast iron circulating pump is suitable for this closed loop system. This pump is connected to the supply side of the collector. The maintenance of pressure in the system (12 psi) and the use of as few fitting as possible, will de-

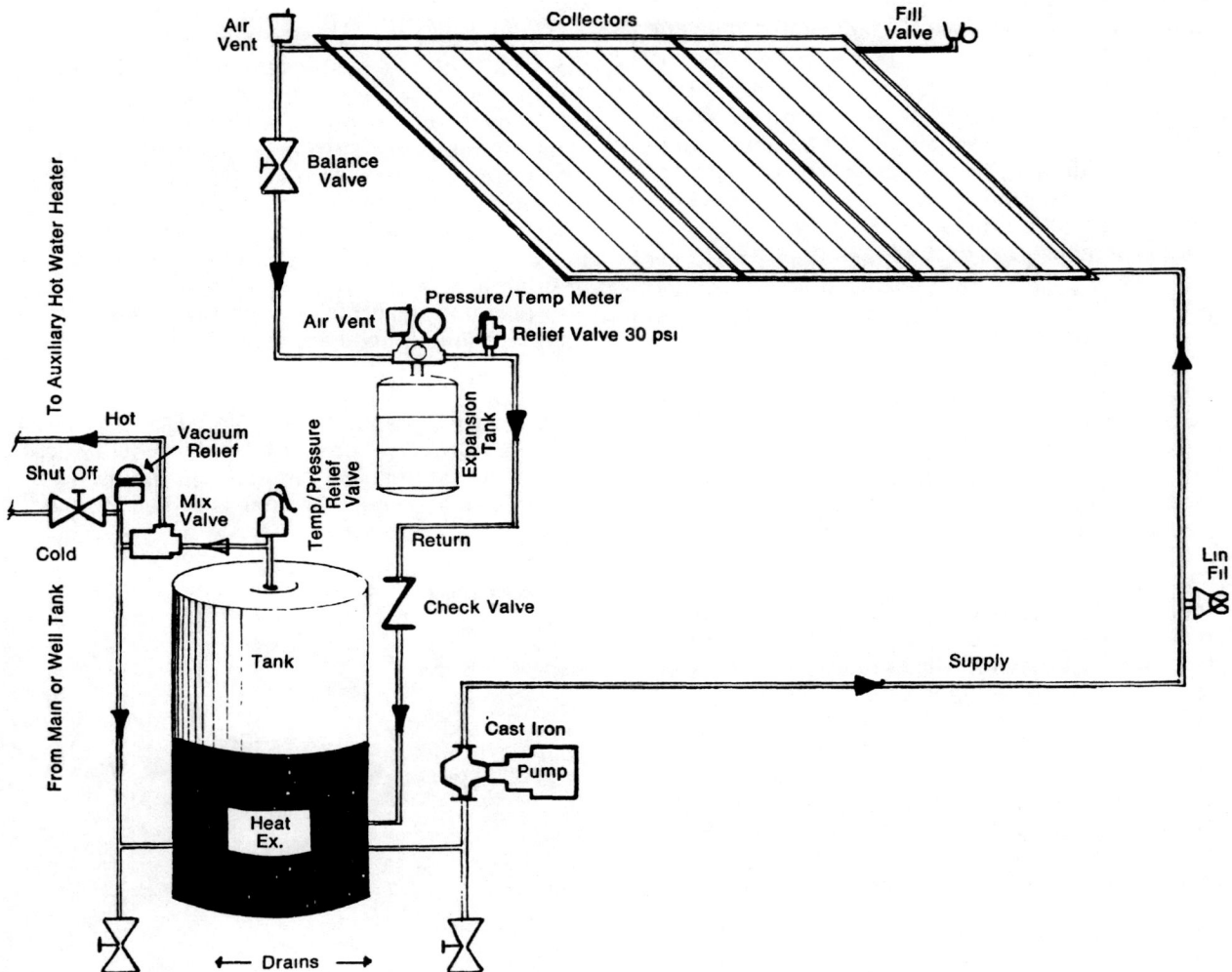

Fig. 9-41. Typical closed loop domestic water heating piping.

crease the load on the pump.

Air Scoop/Expansion Tank. This allows the water/glycol in the system to thermally expand within the limits of maximum pressure rating of the system (pressure relief valve set at 30 psi). An air scoop and vent can be placed at this point in the system if desired or if not venting can take place, at the top of the system.

Vent. This vent in the top of the air scoop vents the air trapped there.

Pressure. A temperature indicator is used to indicate system pressure and temperature. Pressure regulation is important in a closed system to avoid leaks of the glycol.

Relief Valve. A 30 psi relief valve, placed either in the side of the air scoop or a tee in the return line next to the expansion tank, allows the water and glycol to escape if maximum pressure is exceeded.

Flow Control. The number of panels will determine the proper flow rate. Balancing valves are adjustable over the range required for any number of panels. Limiting flow assures proper control of velocity through the collectors and maintains optimum collector efficiency. This valve must be copper or bronze, *not* cast iron.

Check Valve. A check valve must be inserted to stop thermosiphoning when tank temperatures are higher than collector temperatures.

Temperature Pressure Relief. A temperature, pressure relief valve limited to 125 psi should be installed above the hot water outlet side of the tank. 125 psi is used because of limit imposed by the components in the system. The temperature/pressure relief drain must not be valved in any way and must run down to the side of the tank to avoid scalding and water damage.

Mixing Valve. The upper temperature limit of the tank is controlled by the temperature and pressure relief valve—usually 210°F. Although tank temperatures this high are unlikely, the need to control outlet temperature for economy and safety is provided for by a mixing valve. The common setting for this valve is 140°F.

Vacuum Relief. A vacuum relief valve, installed above the cold water inlet of the tank, alleviates a vacuum condition which can collapse the tank.

Air Vent. An air vent is necessary at the highest point in the system (the collectors) to allow trapped air to escape.

Copper Water Lines. All piping runs can be made in ½" copper. Certain fittings and valves will need to be adapted up to ¾" or 1". If runs are long, ¾" lines can be used to reduce head loss

(see pump section). Use 95/5 solder for sweat connections. Keep elbows to a minimum. Use 125 psi rated gate valves for shut-off of cold water supply or tank drain. Dielectric unions should be installed between any ferrous and copper lines. All collector piping must be run *reverse/return*, supply and return at opposite ends of the array. The following list of valves and fittings may be required by local code or desired by customer or installer:

1. check valve and shut-off valve for cold water supply.

2. drain valve for system,

Fill Valve. A fill valve at the top of the system is used to manually fill the lines with antifreeze. The line fill valve is used to introduce a small amount of water to pressurize the system.

Antifreeze. Propylene or ethylene glycol is introduced into the system through the fill valve. The mixture of water to antifreeze should be determined by the chart applied with the materials. The total volume of the system is found by filling and pressurizing the system to 12 psi with water, draining the water into a container and measuring total system capacity. Determine water to glycol mix according to this total. The water introduced to bring the system up to pressure must be taken into account when making the glycol/water mixture. Be sure all lines are vented when system is filled.

Drain-Down System.

(SMACNA Sections 19.1 through 19.6) Typical recommendations on installing drain down systems as shown in Figure 9-44 are as follows:

Fig. 9-42. Fastening treated 2 x 4 standoffs prior to mounting collectors for domestic water heating.

Collectors. Mount at determined angle and pitch, as close to tank as possible. Pitch collectors and supply and return piping so that lines will drain.

Tank. Support tank off floor using concrete blocks if possible to allow for draining off and corrosion protection if basement flooding should occur. The tank should be near the existing hot water heater since the feed (supply line) to this tank is interupting the supply line to the existing system. This will allow the solar system to act in a "preheat" mode as well as supply 100% capacity. A conventional water heater without a fuel system is a very practical solution.

(Note: A tankless coil water heating system maintains boiler water heat independent of coil heat! This will allow only a small decrease in oil or gas usage. Consult the boiler manufacturer as to modifications which will allow sensing of coil temperature not minimum boiler water temperature.)

Pump. A bronze or stainless steel, *not* a cast iron pump, has to be used on a potable water system. These pumps are connected to the supply side of the collector system. The maintenance of pressure in the system and the use of as few fittings as possible will decrease the load on the pump. The low gpm of solar systems will allow the use of 1/12th to 1/25th horsepower pumps in most installations.

2-Way Solenoid Drain Valve. This is a normally open valve which will allow the supply and return lines of the collector loop to drain when the frost sensor shuts off power to the valve. Follow instructions supplied with the valve for proper installation.

Fig. 9-43. Three panels for water heating are mounted on stand offs over existing roof.

Check Valve. Stops "short-circuiting" flow thru the drain loop. Installed so drain-back of the return line from collector will flow thru motorized drain valve.

2-Way Zone Valve. Acts as a check valve between pump and drain valve. Stops flow from tank thru drain.

Flow Control. The number of panels will determine the proper flow rate. Balancing valves are adjustable over the range required for any number of panels. Limiting flow assures proper control of velocity through the collectors and maintains optimum collector efficiency. This valve must be copper or bronze, *not* cast iron.

Check Valve. Installed between flow control and tank. Stops flow from tank through drain.

Temperature & Pressure Relief. A temperature, pressure relief valve limited to 125 psi should be installed above the hot water outlet side of the tank. 125 psi is used because of limit imposed by the components in the system. The temperature/pressure relief drain must not be valved in any way and must run down to the side of the tank to avoid scalding and water damage.

Mixing Valve. The upper temperature limit of the tank is controlled by the temperature and pressure relief valve—usually 210°F. Although tank temperatures this high are unlikely, the need to control outlet temperature for economy and safety is provided for by a mixing valve. The common setting for this valve is 140°F.

Vacuum Relief. A vacuum relief valve, installed above the cold water inlet of the tank, alleviates a vacuum condition which can collapse the tank.

Vacuum Relief (Collectors). Used to allow positive venting through the collector piping when in the drain-down mode.

Air Vent. In this drain-down system, a float type air vent serves two functions.

1. The vent, when placed at the highest point of the system, eliminates the air from the system.

2. The unit allows air to escape while the system is refilling.

Be sure that the vent cap is loosened according to instructions.

Copper Water Lines. All piping runs can be made in ½" copper. Certain fittings and valves will need to be adapted up to ¾" or 1". If runs are long, ¾" lines can be used to reduce head loss (see pump section). Use 95/5 solder for sweat connections. Keep elbows to a minimum. Use 125 psi rated gate valves for shut-off of cold water supply or tank drain. Dielectric unions should be installed between any ferrous and copper lines. All collector piping must be run "reverse/return," supply and return at opposite ends of array. The following list of valves and fittings may be required by local code or desired by customer or installer:

1. Check valve and shut-off valve for cold water supply.

2. Drain valve for system.

Air Systems

(SMACNA Sections 19.1 through 19.6) Figure 9-45 shows the plumbing arrangement for an air-to-liquid solar water systems. The water loop is in-

stalled in a manner similar to liquid to liquid systems.

Figure 9-46 shows the assembly of the air to water heat exchanger. Large insulated ducts connect the heat exchanger to the solar collectors. While thee is no danger of freezeup in the collector loop circuit, there is a possibility of freezeup within the heat exchanger if it is installed in unheated spaces or if cold air from collectors (on sunless, cold days) is moved across the heat exchanger without the water circulating pump in operation.

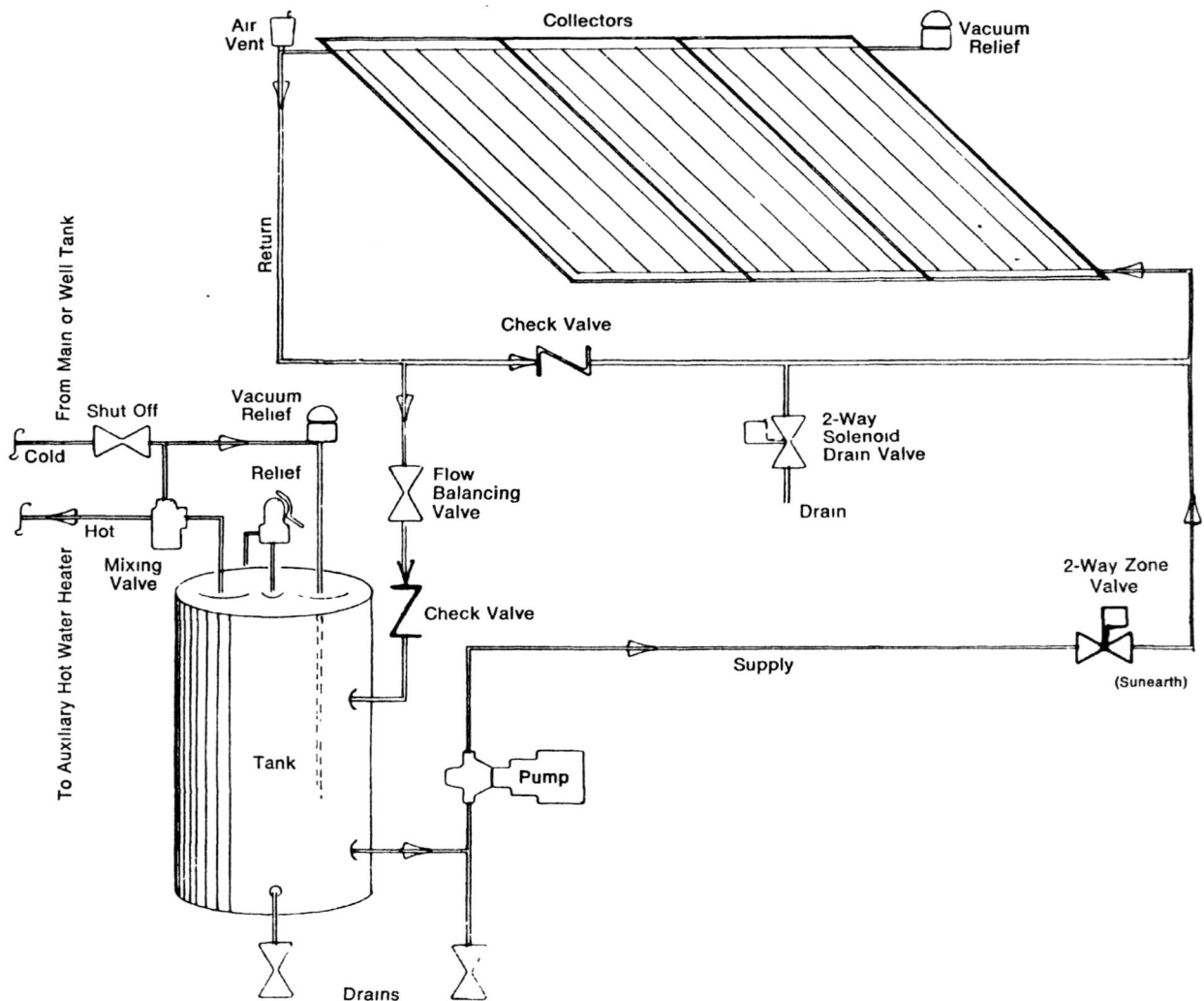

Fig. 9-44. Typical drain down water heating piping.

Although it is a remote possibility, here are some guidelines to further reduce the danger of freezeup when a DHW unit cannot be placed in a warm area:

1. Insulate the exterior of unit with 1″ thick *rigid* insulation, 2 lbs. density per square foot (R-4 to R-7).

2. Insulate *all* piping.

3. Install a "freeze stat" so that sensing bulb is on bottom row of copper tubing inside of the heat exchanger unit. Freeze stat can ener-gize either the circulating pump or heater tapes when the temperature drops.

4. Install a drain-down system using solenoid valves that will be "Fail-Safe" in an open mode for system draining.

5. Install drain pan directly under unit and pro-vide drain according to local building codes.

6. Unit and attached piping *must* be protected from drafts or cold ambient air.

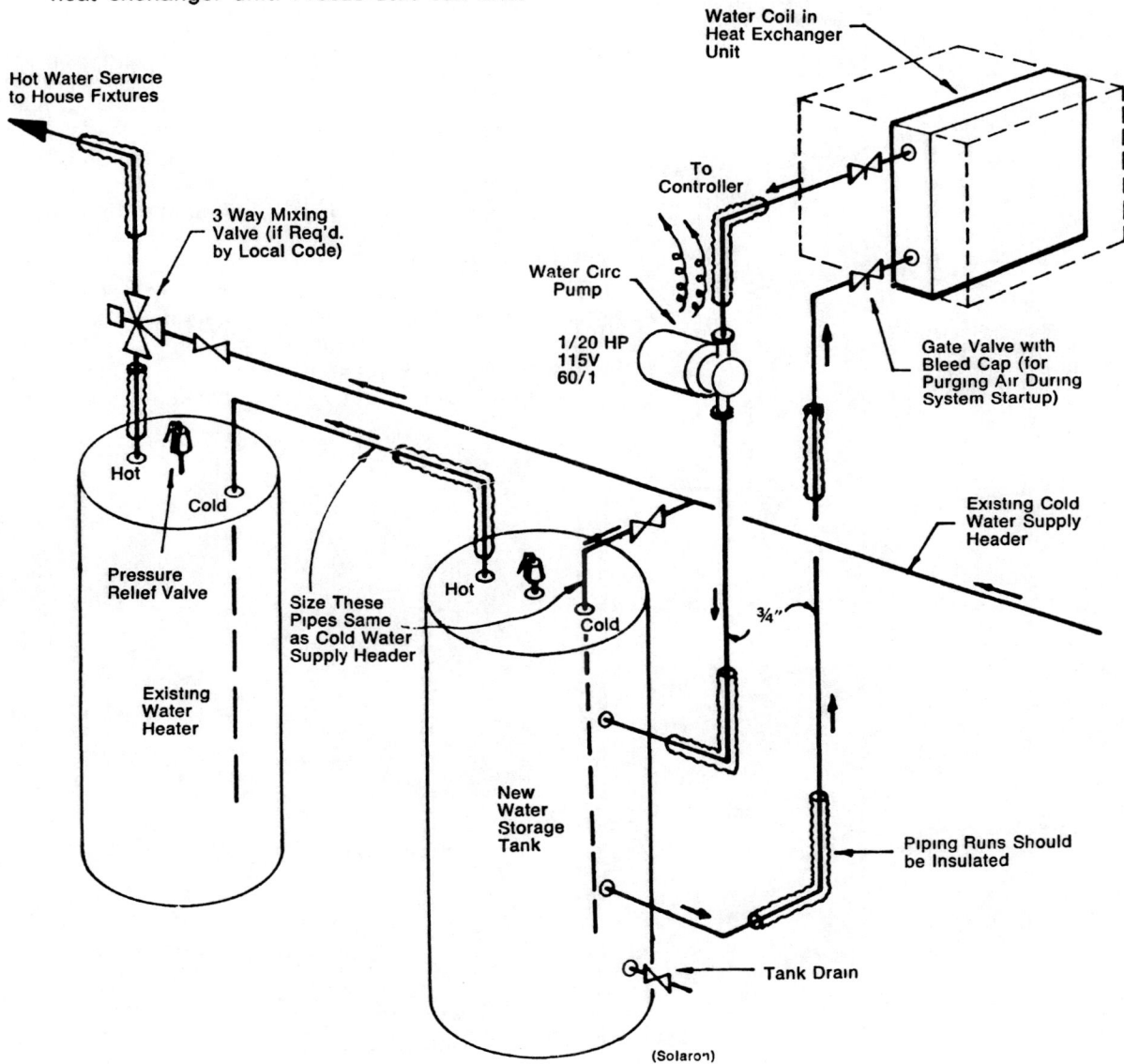

Fig. 9-45. Air system domestic water heating piping arrangement.

7. Install backdraft dampers in both inlet and outlet ducts where penetrators between cold and warm areas are made.

RETROFITTING

Retrofit is a term used to describe the practice of adding solar heat to an existing heating system. The decision to retrofit must be made on the basis of economics. What will the cost be for installing an efficient heating system? Can heat loads be calculated and reduced as necessary? Will the structure support the anticipated additional weight? Is there space for the various components?

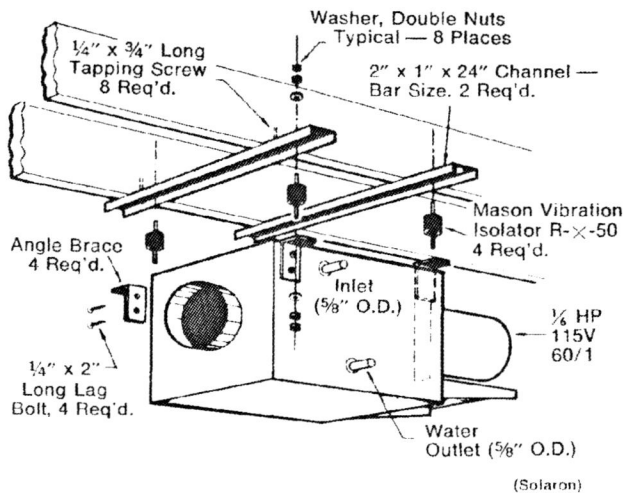

Fig. 9-46. Mounting air to liquid heat exhanger.

Fig. 9-47. Collectors retrofitted on existing roof using existing slope even though it is less than ideal. Additional collector area makes up for the reduced performance.

Collector Array. Most existing structures have been designed with roof pitches that are less than the ideal angle required for flat plate collectors to optimize collection of solar energy. The very common 4/12 pitch roof for example, provides about an 18-½ degree inclination. While racks and stand offs can be used, it's conceivable that the collector area required would make a residential application impractical—certainly unsightly. One solution is to accept the angle provide by the roof slope and increase the amount of collector area needed as a result of lost efficiency.

Another approach is to use a south vertical collector along a wall. Ground locations are possible when the problems of breakage, shadows and ducting are solved.

Still another solution is the erection of adjacent structures that may serve not only as a mounting platform for collectors but other needs as well; porch, carport, etc.

As with new construction, the addition of insulation to the structure to minimize the collector area required is quite obviously the first step.

Pipes and Ducts. Heat transfer mediums will force the loss of some useable space in the structure. A corner of a room or closet may be the area

(Research Products)

Fig. 9-48. Adjacent structure used to mount collectors.

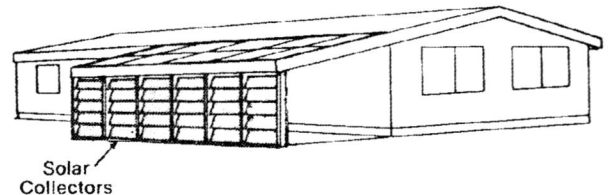

Fig. 9-49. Vertical porch wall used to mount collectors.

where pipes or ducts from the roof to the basement can be placed, insulated, covered with dry wall, and redecorated. Existing joists, studs, sewer lines, and other structural features may result in additional duct bends and pipe elbows that restrict fluid flow and ultimately reduce the efficiency of the system.

Heat Storage Units. In new construction, heat storage units are set during or immediately following basement construction. Retrofitting creates a different set of problems. How can the water tank or pebble-bed be placed in the basement? What are the alternatives if interior placement is not feasible?

Pebble-beds could be located underground beside the basement of the structure. Forms could be placed and footings and walls poured or concrete blocks used. Then, openings in the basement wall would have to be made for connection to the collectors and auxiliary heat source. Pebble-beds could be fabricated with wall studs and plywood in the basement. After pouring a pad for a foundation, the walls could be made of ¾ inch plywood on each side of 2 × 4's about six feet long. Fiberglass batt insulation can be placed between the studs. After setting the wooden storage unit in place, several metal reinforcing bands

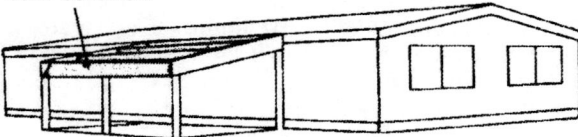

Fig. 9-50. Carport roof used to mount collectors.

(Sun Unlimited Research Corp.)

Fig. 9-51. Maximum insulation must be installed in a solar application to assure economical operation as well as practical component sizing.

should be placed across the entire assembly for added strength. A top can be made and attached after the rocks and sensor(s) are put inside.

Heat storage units for liquids can be placed on the ground in mild climates, buried, or set in a covered concrete pit beside the structure. Here again, insulation is a problem. Another possibility would be to remove a portion of the basement wall and place the tank inside on pre-poured footings. Still another alternative would be to bolt small sheet steel sections together to fabricate a tank inside the basement.

Using an attached garage could be an equitable and the least expensive solution to heat storage unit placement. This would be particularly true if the space was relatively close to the collectors and the auxiliary heating system.

Heat exchangers and air handling modules will be less of a problem to get into the building. But it is noteworthy that they will occupy a considerable amount of space. The air handler will require some customized sheet metal ductwork for connecting the solar heat unit to the auxiliary heat unit.

The other phases of solar heating installation are the same for retrofitting as they are for new construction. These would include placing sensors, controls, and insulation in the appropriate locations according to approved practices.

INSULATION

Insulation is a critical factor in an efficient heating system for any structure. Any discussion of insulation must include the system used by the manufacturers to measure the heat transfer qualities of their products. This refers to the conductivity in Btu's per hour for each square foot [Btu/(ft²·h)] of the insulating per degree temperature difference for each inch of insulation thickness. The *higher* the K value, the *less* effective the insulation.

It is recognized that the design and geographic location of a structure will determine the actual insulation demands.

A point from which to *begin* estimating the amount of insulation needed in currently designed structures is as follows:

Attic:	R-20 to R-30.
Sidewalls:	3½″ (or full thickness)
Windows:	Double pane (thermopane or storm windows).
Doors:	1½″ wood with storm door.
Elsewhere:	Vapor barrier, weatherstripping, and caulking to minimize air leakage.

It should be recognized that these specifications are minimal and they will vary depending on geographical location and construction design.

The whole concept of adequate insulation is based on steps which can be taken to counteract heat loss. Loss of heat from a structure is one of the significant factors used when determining the percentage of the average total heat load that can be designed into a solar heating system. At first thought this, refers to walls, ceilings, windows, doors, and fireplace chimneys. However, there are some additional areas of concern for a solar heating system. For example, the back and sides of the solar collector must be well insulated. This task is accomplished during the manufacture of the component. In addition, insulated heat transfer pipes and ducts must transmit heat efficiently to well insulated heat storage and distribution equipment. Materials used to insulate the solar heating system should have: (1) low thermal conductance; (2) high resilience; (3) resistance to fire, insects, weather, mildew, etc.; (4) a high melting point; (5) low bulk density; and (6) a surface that is resistant to abrasion. Also, insulation materials should be economical to install.

Fiberglass, for example, is one of the materials that meets these criteria. The extent to which fiberglass meets a specific criteria will vary; however, this insulating material has many uses. *Liquid* storage tanks (space and water heating) should have 6″ or more of insulation. All heat transfer mediums (liquid and air) should be covered with at least 1″ of insulation (R-4 to R-7). In some special cases, pipe, duct, and tank insulation can be reduced or eliminated when the transfer mediums are within the space designed to be heated.

The pipes for liquids and round ducts for air should be insulated with pre-formed materials which can be slipped over the round pipes and sealed. Pre-formed or molded coverings for elbows and bends should also be used. Some soft flat materials such as fiberglass could be shaped to fit around a pipe but this is not an acceptable insulating practice. The reason is that the effectiveness of the insulation is reduced when the air cells are crushed wherever the material is creased. The same inefficiency results from squeezing four inches insulation into a narrower space. Insulation should also cover pipes and ducts where they pass through floors, walls, and ceilings. Duct and pipe changer straps should be placed after the insulation has been installed.

Insulation is an extremely important factor to the solar heating system. Economics would dictate that it is more important to design and install a 50% heat load solar heating system that approaches 100% efficiency rather than a 100% system that is 50% efficient. (For more information on insulation, refer to SMACNA Section 1.1-1.7 and 9.1-9.6).

SUMMARY

It is quite obvious that there are factors to consider when installing a solar assisted heating system in new construction or retrofitting it into an existing building. Adequate insulation will have to be provided. Space will have to be allocated to housing the necessary parts of the system. There are problems of supporting the collectors on the roof regardless of the pitch. Choices must be made between the liquid or air system and what percentage of the total heat load can be effectively planned for a given geographical location.

Regardless of the decisions about how the system is to be used, it must be installed and maintained properly or the life of the system will be significantly reduced. Specific steps required to properly balance and "tune" the system when it is installed are very similar to troubleshooting, repairing, and "retuning" services. These activities are presented in the next lesson on servicing.

SERVICING

A solar heating system must be serviced to operate efficiently and effectively. The architect, heating engineer, heating system installer, and the owner of the structure must have an understanding of the operational principles of solar heating systems.

There are three kinds of services performed on a solar heating system. *First* is the installation start-up, during which the system becomes operational and is tuned for optimum efficiency. The *second* phase of servicing involves periodic maintenance of the system throughout the lifetime of equipment and, *third*, troubleshooting or emergency service.

All servicing must be done *safely.* When working around solar heating systems, care must be taken to guard against burns and electrical shock. Collectors are capable of reaching heat levels of up to 400°F. The possibility for electrical shock is the greatest when contact is made with electrical terminals or the wiring, particularly if the area underfoot is wet as a result of a liquid leak or other causes. One of the reasons for using 24 volt circuitry for sensors, relays, and thermostats is that the low voltage is safer.

INSTALLATION START-UP

General Information

The responsibilities of the installer and designer for the efficient operation of the solar heating system are closely related. Theoretically, the designer should take enough time to plan and specify every part of the system down to the location and size of the last screw. In actual practice, however, the designer passes a good deal of design responsibility on to the person who does the installation work. It is assumed that the installer is a skilled mechanic who recognizes and practices standard detail work that results in a good installation. The designer usually takes the responsibility for selecting the proper size collectors and equipment, sizing pipes and ductwork, selecting and locating supply registers and return grilles, and designing the control system. The installer assumes all other work in connection with the system.

Trouble which cannot be traced to a defective part, maladjustment, or installation fault may ultimately be the responsibility of the designer. No field correction will add output to the collector or other components which were selected without sufficient capacity in the first place. However, it is sometimes possible for the installer to reduce building head load demands and solve the problem.

Duct systems that are too small can often be corrected by adding a branch or two. Distribution which does not provide satisfaction because of drafts or uneven heating may often be improved by relocating a supply register or return grille.

Controls may fail to function properly because of poor choice, location, or lack of owner understanding of their purpose. These are design faults which can usually be corrected on site.

Solar equipment is manufactured within very narrow operational tolerances. There are many models, types, and sizes of units to meet practically every application requirement. However, from the service viewpoint, one cannot always assume that the proper equipment selection was made to fit the application. Neither can it be assumed that the operating conditions, as one views the installation, are the same as the original design conditions.

It is important that customers are questioned regarding a change of operating conditions. Also, it would be helpful if they understood the effects

Fig. 10-1. Don't touch anything inside a master control panel or wiring compartment of a solar heating system until you have checked across every power terminal and each terminal to ground to be positive all power is off. Use test meters and test lamps with well insulated probes and leads to test circuits and components. And use JUMPERS with great discretion. Components can be readily damaged by misconnected jumper leads. Be alert and cautious at all times.

of these changes on the operation of the equipment to facilitate service and operation.

Manufacturers have added *safety devices* to protect equipment from operating beyond design limitations as well as providing protection when a component fails. Since the prime source of power is electricity, safety devices are incorporated in the electrical circuits to interrupt the flow of current should danger threaten the system. For example, an electrical device may wear out or an electrical component may break down because its limits have been exceeded. Either problem would cause the equipment to stop operating thus preventing more serious damage to the system.

Start-Up Liquid System

Start-up begins after the various components have been set in place, the system leak tested and electrical controls connected. Start-up involves charging the system, testing control modes, and taking measurements of fluid flow rates, temperatures, electrical power, input, etc.

Most collector manufacturers provide a recommended fill procedure. It usually involves the use of an auxiliary "charging" pump, hose connections, and perhaps a special container if a heat transfer liquid is involved. One example is shown in Figure 10-4.

As the system is filled, various drain valves must be closed and the air vents checked repeatedly throughout the filling process to insure they are functioning and not jammed by some foreign materials. (Remember, the system has already been leak tested and flushed during installation.)

All the air will not be vented on the initial fill since some air remains "in suspension" in the heat transfer fluid and will not separate until the liquid is heated. This is typical for even ordinary hydronic

(Hydronics Institute)

Fig. 10-3. Following manufacturers' instructions is absolutely essential for proper installation and start up.

Fig. 10-2. Designer takes responsibility for sizing and selecting equipment and related components.

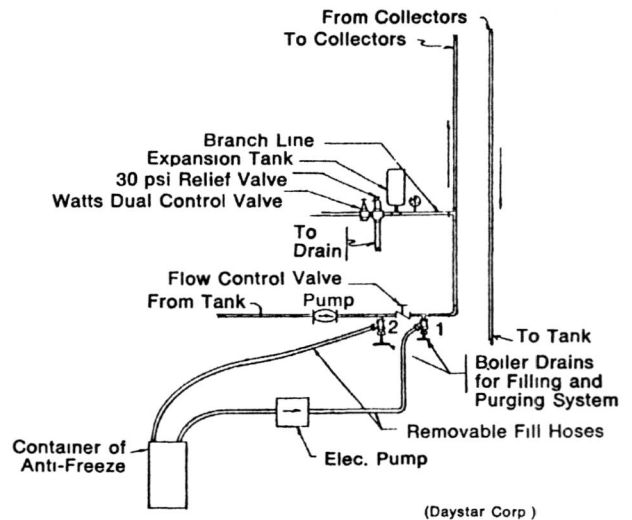

(Daystar Corp.)

Fig. 10-4. A fill system to "charge" a collector loop.

heating systems. However, to insure all air is removed, the serviceperson can manually open all bleed valves every day for the first several days of operation. Late afternoon purges are to be preferred since any air in the system will rise to the upper part of the system near the collector discharge piping.

After all of the air has been vented, it is desirable to cap any drain valves to prevent accidental draindown of the system.

Many differential controllers have a three position switch—*on, off,* and *automatic.* The *on* position can be used for test purposes to determine if the pump as well as various control valves are operating. Upon completion of this basic test, the switch must be placed in the *automatic* position. Controller manufacturers usually provide testout procedures. One example is shown in Figure 10-5.

Sensors are the components from which the controller receives an electrical signal. Therefore, malfunctions may *not* be the fault of the control-

ler but of the sensor. Correction of the problem can be accomplished by checking the sensor as recommended in Figure 10-6

Once the collector temperature increases to the controller set point and pump operation in the automatic mode has begun, an operational check should be completed to record the necessary data for a given system. This list should be developed by the contractor and his component suppliers to meet the specific characteristics of the proprietary design for future servicing points of reference. Figure 10-7 is a sample of such a checklist.

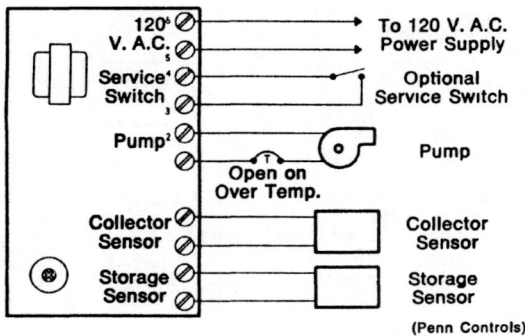

Typical Wiring Diagram

If circulating device (pump or blower) fails to energize when conditions indicate it should be running, proceed as follows

1. Use a thermometer and check to be sure the proper differential does exist.
2. Check for proper voltage (120 V. A.C) supply to terminals 5 and 6.
3. If Steps 1 and 2 check all right, disconnect the collector sensor leads from the controller. This stimulates a very high collector panel temperature and the pump (blower) should energize. Another way to test this function would be to reconnect the collector sensor and short circuit the storage sensor. This simulates an extremely low storage temperature and the pump (blower) should energize.
   If this step energizes the pump, a defective collector sensor and/or storage sensor is indicated. Refer to sensor checkout instructions. (See figure 10-6).
4. If the sensors are operational per Step 3, short circuit terminals 3 and 4 for the manual override and if the pump (blower) energizes, a defective manual override switch is indicated. Check wiring to switch Replace the switch if bad.

Fig. 10-5. Checking differential controller.

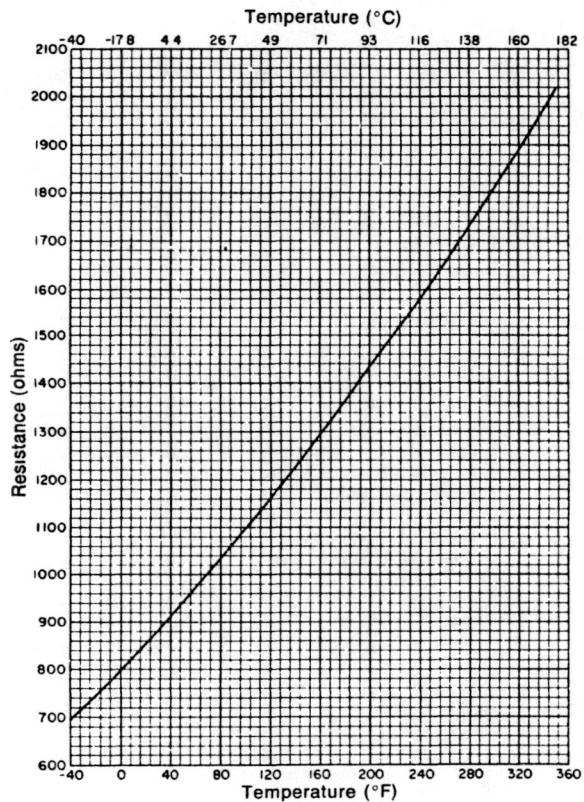

Temperature vs Resistance Graph

If faulty sensor(s) is suspected proceed as follows.

1 Disconnect sensor wires.
2. Measure temperature at sensor.
3 Measure resistance of sensor with an ohmmeter. An open or short measurement indicates a bad sensor.
4. Check temperature measured in Step 2 and resistance measured in Step 3 against the curve above
5 Replace sensor if it is defective.

Fig. 10-6. Sensor checkout.

Start-Up—Air Systems

Generally speaking, the start-up of air systems is considerably easier than liquid systems since the fill/venting procedure is unnecessary. In all other respects, control checks, and temperature and power measurements are essentially the same. *Before* actual start-up, follow these *precautions:*

1. Check for proper mounting of belt-drive motor.

2. Check belt tension.

3. Check pulleys for tightness on shafts.

4. Remove all tools, materials, etc. from inside the unit.

5. Check auxiliary heating unit per manufacturers' recommendations.

6. Activate electrical power to air handling unit and controller.

7. Check rotation of air handler blower.

8. Activate electrical power to auxiliary heating unit.

9. Secure all access doors.

10. Check operation of all components and systems per control instructions.

Once the system is operating and there are no immediately obvious emergencies, the *operational checklist* as outlined in Figure 10-7 can be completed.

System Balancing

In the operational checklist of Figure 10-7, balancing of both air and liquid flows is essential to the proper operation of the solar assisted heating system. While many conventional residential hydronic and warm air systems are not thoroughly and accurately balanced, it is necessary that com-

OPERATIONAL CHECKLIST DATA

1. Date of recording _____

2. Atmospheric conditions _____
   (sunny, cloudy, etc.)

3. Ambient temperature _____°F

4. Thermostat settings
   4.1 Heated space _____°F
   4.2 DHW heater _____°F

5. Collector
   5.1 Absorber plate temperature _____°F
   5.2 Collector Outlet temperature _____°F
   5.3 Collector Inlet temperature _____°F
   5.4 Condition of transparent covers _____
   5.5 Condition of cover mountings _____
   5.6 Condition of absorber plate _____
   5.7 Condition of flashing _____
   5.8 Evidence of leakage _____
   5.9 Balance of manifold _____
   5.10 Emergency purging components _____
   5.11 Pressure drop (manometer reading) _____

6. Pipes and Ducts
   6.1 Condition of insulation _____
   6.2 Evidence of leakage _____

7. Heat storage unit
   7.1 Top temperature _____°F
   7.2 Middle temperature _____°F
   7.3 Bottom temperature _____°F
   7.4 Evidence of leakage _____
   7.5 Condition of insulation _____

8. Fluids
   8.1 Cleanliness
       8.1.1 Filters in liquid system _____
       8.1.2 Filters in air system _____
   8.2 Antifreeze (hydrometer test) _____°F
   8.3 Acidity _____
   8.4 Evidence of corrosion _____

9. Heat exchanger (liquid closed loop or air handler)
   9.1 Collector inlet _____°F
   9.2 Storage outlet _____°F
   9.3 Auxiliary heater outlet _____°F
   9.4 DHW preheat _____°F
   9.5 DHW cold water supply _____°F
   9.6 Pool preheat _____°F
   9.7 Pool water-return _____°F

10. Dampers
    10.1 Cleanliness _____
    10.2 Condition of weatherstripping _____
    10.3 Evidence of leakage _____
    10.4 Motor(s) operation _____

11. Valves
    11.1 Evidence of leakage _____
    11.2 Evidence of corrosion _____
    11.3 Solenoid effectiveness _____
    11.4 Cleanliness of filter screens (if present) _____

12. Auxiliary heating systems
    12.1 Evidence of leakage _____
    12.2 Electric heat
         12.2.1 Condition of lead-in wire _____
         12.2.2 Condition of heat element _____
    12.3 Liquid or gaseous fuel
         12.3.1 Evidence of leakage _____
         12.3.2 Condition of upper limit switch _____
         12.3.3 Condition of lower limit switch _____
         12.3.4 Condition of automatic valve _____
         12.3.5 Condition of pilot flame _____
         12.3.6 Condition of burner flame _____
         12.3.7 Condition of electrical controls _____
         12.3.8 Condition of blower _____
         12.3.9 Condition of air-return filter _____
         12.3.10 Condition of sediment filter (oil) _____

Fig. 10-7. Sample operational check list.

plete balancing procedures be followed using accurate testing and balancing instruments. Each system should, therefore, be planned from the very outset to ease the balancing task by providing for pressure and temperature measurement stations. Before any attempt is made to balance a system, time should be spent to prepare. Here are several items which should consider *before* starting the job:

1. Inspect the entire job plans and decide on a specific approach to balance the system. By reviewing the plans and specifications, a contractor can determine a definite step-by-step procedure. At this time, it can also be determined if there are any variations between design drawings and shop fabricated drawings.

2. Make up a complete set of work sheets in advance. Properly organized, these data sheets can produce a time-saving record of all measurements made as the system is being balanced.

3. Have available register manufacturers' recommendations for measuring air quantities at the diffusers and registers. Then prepare a list of needed air flow measurement factors (K values).

4. Have available pump and fan performance data as well as collector pressure drop information as supplied by component manufacturers.

5. Plan balancing procedures so that the back dampers provided with the diffusers or registers will be used only as the *final trimming* of the air distribution system. Do *not* do this for *initial system balance* since it often leads to complaints of excessive whistling noise and possible poor air distribution.

Instruments for System Balancing

There are many air balancing instruments available and the question is often asked, "Which ones are needed to do a good job?" Experience suggests that the following instruments are needed:

1. A deflecting vane anemometer, such as the Alnor "Velometer," for measuring air velocity.

2. A rotating vane anemometer and stop watch—also for measuring velocity.

3. A Pitot tube and connecting hose to sense static or total pressure.

4. Various pressure gauges such as a manometer, U-tube, slope gauge, and magnehelic gauge for indicating pressures.

5. Thermometer for temperature measurements at various duct locations.

6. Tachometer for measuring fan rpm.

7. Volt-amp meter for checking the fan motor voltage and amperage.

The importance of an instrument that is *properly* calibrated cannot be overemphasized. The instrument should always be checked before balancing a job. With this in mind, these precautions should be followed:

1. Always follow the instrument manufacturer's recommendations for checking the calibration of the instrument. Most instruments are very delicate and if they are dropped or bumped excessively, calibration can be changed. Using an *out-of-calibration* instrument can only result in a poorly balanced system.

2. When possible, the same instrument should be used for the entire job. More instruments multiply the danger of calibration errors.

3. If more than one instrument has to be used, a check should be made to compare how close they read. Variations between instruments should not be greater than plus or minus 5 percent.

For a thorough explanation of balancing procedures, refer to SMACNA's *Testing, Balancing, and Adjusting of Environmental Systems* manual.

PERIODIC MAINTENANCE

Correct maintenance is perhaps even more critical for solar systems than for conventional systems, and the owner must be aware of this. Contractors might offer a yearly "service package" for those consumers who do not wish to assume maintenance responsibilities. The items of maintenance which *must* be emphasized are:

1. Air filters—change monthly for the first three months and quarterly thereafter.

2. Liquid filters—change after the first month and annually thereafter.

3. Antifreeze solution—check annually, change when color indicator changes color.

4. Blower drive belts—check annually and change if worn.

5. Pumps—lubricate at fixed intervals suggested by manufacturer.

6. Liquid level—check weekly for first month

and quarterly thereafter. If liquid level drops, check for leaks. Also, check any time the electrical power has gone off during the day. If low, refill with proper antifreeze solution.

Component Maintenace

Here are specific maintenance steps for each solar component:

Collectors. Collectors are the heart of the solar heating system. Servicing involves troubleshooting and repair of various elements in the collector.

Transparent Covers. Outer covers should be free of dust and dirt, chips, cracks, condensation, etc., and the rubber mounts should not be cracked from the effect of weather and heat. Inner covers should be inspected for the same kinds of physical deterioration.

Absorber Plates. Different inspection practices will be utilized for air or liquid absorber plates. The fluid transfer tubings and ducts for each type of collector should show a similar pressure drop between the inlet and outlet fittings. Pressure drop across each collector is measured with a manometer (Figure 10-9) or by temperature rise.* The fluid levels in the two sides of the loop will be even when the collector is not operating. When fluid is passing through, there will be a difference in the height of the liquid in each tube. An efficiently operating collector will have about ½″ to 1″ difference in liquid levels in the tubes.

The temperature at the top of each collector array should be the same. A heat flow meter, thermometer, or surface pyrometer can be used to measure this characteristic. Differences in temperature can be balanced by opening or closing the flow rate valves at the top (outlet manifold) of each collector or array. If adjustments do not correct outlet temperature differences, there may be corrosion constricting the tube which will have to be removed.

Inspect the absorber sensor. It should be attached firmly to the plate and a small amount of thermoconducting compound should be present between the plate and sensor to increase sensitivity.

Absorber plates are coated with flat black paint or special heat-absorbent selected coatings. The manufacturer's literature should identify the proper touch-up material if the surface peels or gets scratched.

---

*On a clear sunny day (all day), around noon, a temperature rise from 15 to 25 degrees for a liquid loop and 50 to 75 degrees for an air collector would indicate effective operation

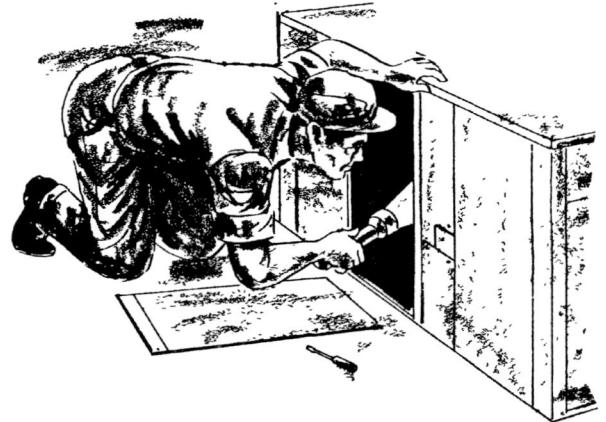

Fig. 10-8. A visual inspection is an important part of system maintenance.

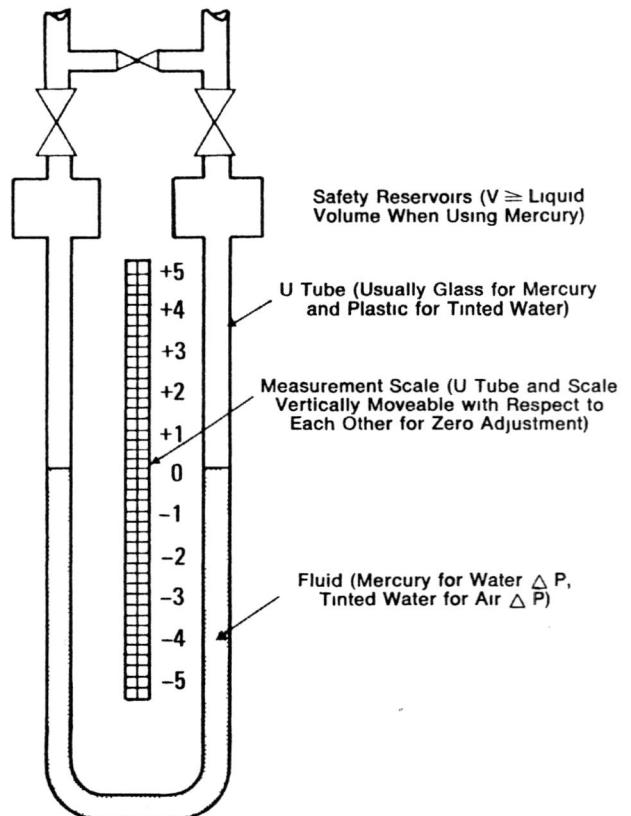

Safety Reservoirs (V ≧ Liquid Volume When Using Mercury)

U Tube (Usually Glass for Mercury and Plastic for Tinted Water)

Measurement Scale (U Tube and Scale Vertically Moveable with Respect to Each Other for Zero Adjustment)

Fluid (Mercury for Water △ P, Tinted Water for Air △ P)

Fig. 10-9. Details of a "U" tube manometer.

Tubing. Tubing on the absorber plate may be loose or have pulled away from the plate. If this happens, heat transfer cement will make the tube stationary and fill in any voids between the tube and the plate. Poor tube-to-plate contact reduces collector efficiency.

Pipes. Liquid transfer pipes should be well insulated. They should be mounted solidly. There should be no visible signs of leakage either from the water in a drain down system or antifreeze in a closed system collector loop.

Ducts. Air ducts should be sealed to prevent leaks. Passing a heat flow meter sensor or simply your hand across the surfaces of each duct section should reveal any hot area thus indicating that the insulation inside has loosened. Disassembly will be necessary to correct the problem.

Heat Storage Units. Storing heat in a pebble-bed or liquid tank should be a rather maintenance-free proposition. Using concrete for a pebble-bed or liquid container poses maintenance problems if it cracks because of poor quality control in the batch of concrete or inadequate footings upon which it is placed. Water backing up from a clogged sewer or inadequate surface water handling systems (resulting from a hard rain) could flood the pebble-bed and cause a difficult rock cleaning problem. If the lower opening in the concrete container is level with the bottom of the pebble-bed, drainage of water when cleaning the rocks can be handled easily.

There are problems with making water tight concrete-to-pipe seals for liquid systems. Silicon can be used under a pipe flange when it is bolted to the container.

Fewer problems exist with glass or stone lined steel tanks for water storage. The unit is maintenance-free unless the water can make contact with the steel and cause harmful rusting.

A totally fiberglass tank would be as near maintenance-free as could be expected. (Remember temperature limitations, however.)

Inspection of the steel or fiberglass water tank insulation will reveal heat loss sources if the covering is split or loose. This problem can be corrected with special tapes and adhesive materials manufactured for that purpose.

Fluid (Liquid). Maintaining the liquids to the proper specifications has two very important functions. Unless an adequate amount of antifreeze is used, extensive damage to the heating system and structure can result from freezing. Care must be taken to check the antifreeze concentration regularly.

Corrosion, which causes metal deterioration, especially iron and aluminum, can be costly. Corrosion inhibitors should be added to the system as needed. The iron neutralizing material in the getter column should be checked and replaced when it decomposes.

Fluid (Air). Contamination of air can be controlled by filtering. Filters located near the pebble-bed in the air return from the heated space need the same kind of attention as those in a conventional heating system.

Air Handler (Air System Heat Exchanger). This units contains electrical controls, plumbing connections (for DHW), air diverting dampers (gravity and/or motorized), a blower and a blower motor. Removal of a side panel allows access to the blower and motor for oiling. Damper adjustments, as well as the rubber seals on the louvers, can be inspected and repaired or adjusted as needed. Water leaks from the DHW coil can be located easily, but additional disassembly may be necessary to make repairs.

A discussion of the electronic controls will be taken up in the controls section of this lesson.

Dampers. Dampers located elsewhere in the air system should be maintained with inspection and repairs as noted for the air handler.

Valves. Valves are subject to leakage and corrosion (from electrolysis the same as other components). The ball in the back-flow check valve must be free to move. A valve replacement must be made if the stainless steel ball inside is not free to move. Solenoids or automatically controlled valves will short out, especially if they get wet, and must be replaced. Valves containing screen filters must be disassembled and their filters cleaned periodically. Air-vent valves must be replaced if the float becomes inoperative. Other valves which are installed to prevent over-pressurizing and/or overheating should be checked to determine if they are operational.

Auxiliary Heating System. It is not within the objectives of this lesson to describe maintenance procedures for auxiliary heating system. The operational checklist presented in this lesson should be adequate to describe the maintenance demands of the system.

Domestic Hot Water Heating. Figure 10-10 provides a simple maintenance checklist for both open and closed water heating systems, including maintenance the homeowner can do.

Sensors. Sensors are generally found to be identical as far as their function is concerned. They are all nickel wire wound resistors with a sensitivity to heat that varies their internal electrical resistance. For example, a sensor may have a resistance of 1000 ohms at 72°F. As the sensor warms or cools, the internal resistance will change 3 ohms per °F. A temperature drop of 20°F that would cause a pump to energize will change the resistance of the sensor to 940 ohms.

Sensor mounting is accomplished by different methods. Some are held in place with screws. Others slide into bulb wells. Pipe threads are on still other models which screw directly into the liquid where temperatures are being monitored.

Sensors are easy to replace unless they are immersed in a liquid. Then, the system must be drained until the water level is below the sensor mounting hole.

All systems should be checked at least once a year

A. Drain-down and Antifreeze Systems

☐ Pump — Check any lubrication points and lubricate as necessary.

☐ Air Vent — Check for clean float and seal. The air vent will "sputter" if there is any debris caught in the unit

☐ Collectors — Clean the outer glazing. All acrylic should be cleaned with mild soap and water followed by a water only rinse.

☐ Tank — Drain the tank and flush to remove sediment.

☐ Check all fittings and insulation for leaks.

☐ Re-balance the system with the flow control fitting.

B Drain-down System Only

☐ Check system pressure and its relation to the setting on the pressure switch Adjust if necessary.

☐ Remove a fitting if possible to check for scaling. If any exists, install a water filter — soft water system such as Calgon's "Season Treat."

☐ Check for clean operation of the solenoid valves by unplugging the line from the auxiliary outlet Be sure that the drain runs freely.

C Antifreeze System Only

☐ Drain and flush the collector loop and re-balance the system.

☐ Refill with antifreeze according to instructions. Check antifreeze once a year for ph and scale. If scale is visible or ph is not between 8 to 9.6, replace.

☐ Check system pressure and relief valve in collector loop for proper pressure range

D. Homeowner Responsibility

1. Keep all debris from collector array (such as leaves, branches, etc.).

2 Allow the system to run if you only plan to be away from the house a few days. If you plan to stay away for a period of a month or longer, turn the system off at the control or breaker panel.

3 If a leak develops, shut down the system immediately Call your installer

4. Periodically check the system pressure, fittings and pipe insulation. If no hot water is circulating from the collectors, check the following before calling a serviceman. (1) Be sure all electrical cords are plugged in, (2) Check low voltage wire connection to face plate on control, (3) Check fuse or circuit breaker which services the system, (4) Be sure valves are open (do not tamper with balancing valve).

5 Conserve water! Install flow controls on shower heads, use the dishwasher only when completely full, do not run water unnecessarily

(Sunearth)

Fig. 10-10. Domestic water heating maintnance check list.

Electronic Controls. The solid state electronic controls for a solar heating system are operated by the line voltage (120 VAC) which services them and the sensors which detect temperature differences and cause the pumps or blowers to be energized. The controls are all factory adjusted to the specifications required for the various modes of operation. If a customer needs different temperature settings than those which are established for standard stock controls, the manufacturer will prove the alternate settings within the control.

One of the principal electronic controls is the *differential thermostat.* It's operation was described in a previous section. This control uses two sensors. One can be located in the collector and the other in the heat storage unit. Upon command, the thermostat will energize a pump connected to the relay terminal and cause the fluid to circulate through the collector circuit. A pump with a maximum operating current of up to 10 amps may be used with this unit. Another model of a differential thermostat has two sets of relay contacts. One set operates as indicated above and the other set may be used to energize the pump to protect the open loop circuit from freeze damage. When this condition occurs, warm water from storage is circulated in the collector to keep it from freezing. Another option would be to use the second set of terminals to cause a solenoid operated drain to purge the collector of water in the event of near-freezing temperatures. Still another option would be to switch on the auxiliary heat source if there were not sufficient heat available from the storage unit or developed from the collectors.

A *proportional* control operates much like the differential thermostat. The capability that this control has is to be able to operate a pump or blower at either of two speeds whenever the load demand is high or low. Obviously, a two speed electrical motor must be used with this control. By operating a pump or blower at two different speeds, the need for balance valves to adjust flow rate may be eliminated.

A control, called an *activator,* may serve many purposes. It can drain a collector, power a small heating element added to a collector to prevent freezing, change operations from one storage tank to another, activate the auxiliary heat source, or perform any other function that is desired.

Pump-operating amperages and voltages may be required that exceed the ratings of various controllers for heavy duty circulation systems. The regulators described above can be used, but another relay must be added for the extra current and voltage requirements of the heavy duty pump.

Maintaining motors or other devices with solid state electronic control equipment is relatively simple. *For example,* if a pump does not start when it should, use a thermometer first to verify that the necessary temperature differential exists to activate the unit. *Second,* check for adequate service voltage (for example 120 VAC). *Third,* disconnect the collector lead from the control. This simulates a very high collector temperature and, as a result, the pump or blower should operate. A similar check by shorting out the storage sensor will simulate an extremely cold condition and signal for the pump to start. If the pump does operate at this time, the collector or storage sensor is defective and must be replaced. The *fourth* check is to short out the service switch. Merely wiring around the switch is sufficient. If the pump operates, replace the switch. *Fifth,* if after all of these troubleshooting procedures the pump still does not function, the electronic control unit must be replaced.

Proper installation and maintenance servicing will optimize the operation of a solar heating system. Proper maintenance will extend the efficient and effective life of a solar heating system for the greatest length of time. A much greater return can be realized on the money invested if the proper servicing steps are taken at regular intervals.

TROUBLESHOOTING

The quality of service can made the difference between success and failure in the heating business. The customer expects not only a good installation, but also prompt and efficient service in the event of trouble. Many successful contractors sell their customer on the practical aspects of preventative maintenance. This means an annual inspection of the heating equipment, made during the summer to insure that the system is in good working order at the start of the following winter. This will reduce the probability of an emergency service call in the middle of the winter.

There is no substitute for *experience* when it comes to efficient servicing. No text could possibly cover all service problems.

By adhering to the rules, procedures, and techniques outlined in this lesson, service problems can be isolated to specific faulty components (or faulty design) in a minimum amount of trouble shooting time. Obviously, the importance of understanding and using detailed service instructions provided by the manufacturer of *specific* makes and models of controls and components cannot be overemphasized.

Unless an individual is an extremely experienced serviceperson, the basic approach to troubleshooting has simply got to be by means of a *process of elimination.* Four basic steps are recommended:

1.   The complaint is noted.

2.   The symptoms are determined.

3.   The cause for each symptom is checked.

4.   The trouble is remedied.

Avoid guesswork. Instead rely on a systematic approach to the problem. Of course, not all guesswork can be avoided. The so-called "educated guess" of an experienced service technician can often save time and money. Past experience greatly assists the organized trouble shooting approach.

When a solar heating system fails to perform properly, the underlying cause will usually fall into one of four categories: part failure, improper adjustment, poor construction, or poor design.

Fig. 10-11. Providing emergency service is a very important part of the overall success of a solar installation

Part Failure

A part failure is, perhaps, the easiest malfunction to correct since, once detected, a simple replacement puts the system back into satisfactory operation. Many part failures are quite obvious; for example, a broken belt, a wornout bearing, or an open circuit breaker. Some, however, require considerable skill to detect. In between, there are failures such as a leaking pipe joint, a burned-out motor capacitor, or a defective control. Following the "process of elimination" procedures may be necessary in order to pin-point the "cause" of these problems.

Parts fail for several reasons, all of which can be summarized as follows: (1) defective in manufacture, (2) subjected to conditions beyond their rated capacity, (3) not properly maintained, and (4) wornout from usage.

Modern factory quality control methods used by the leading manufacturers prevent most parts of defective manufacture from getting beyond the factory. To protect the owner from the few that do, one to five year guarantees are often offered. Any part which functions satisfactorily for a year has a good chance of remaining in good operating condition for many more years of service.

Parts which fail due to application beyond their rated capacity include mostly electrical items such as motors and controls. It is possible to over load fan motors by imposing conditions in the field which are beyond their intended service. A fan motor can easily be overloaded by small increases in fan speed since brake horsepower requirements increase as the cube of the speed ratio. Thus, for example, a fan requiring 1 hp at 500 rpm would need 8 hp at 1000 rpm.

Electrical controls are all rated by their respective manufacturers for maximum ampere draw at each operating voltage. When these conditions are exceeded, the life of the electrical components in the control is shortened considerably.

Parts which fail due to improper maintenance can be either mechanical or electrical. Any system having components exposed to outside weather is subject to rust and other types of deterioration due to rain and prolonged sunlight. This latter is particularly hard on parts made of plastics and on electrical insulation.

Any part which requires periodic lubrication is subject to premature failure when regular lubrication maintenance is ignored. Motor and fan bearings fall into this classification.

The fourth classification of parts failure is wear and tear. Every system will be subject to this type of problem. The most vulnerable items are electrical components. However, mechanical components with moving parts, such as bearings and pump, are not far behind. Scheduled lubrication will prolong their life.

Adjustments

A second major reason why a solar system fails to perform properly is, as noted previously, improper adjustment. As compared to a system which does not function at all because of a part failure, a system that is not in proper adjustment may cause the owner great dissatisfaction. The reason is that, in the former case, the owner is sure something is wrong so the service person is called and the condition is corrected. When a system is out of adjustment, however, the condition may develop so gradually that the owner may not be certain that something is wrong until after a period of considerable annoyance.

A system which is out of adjustment may result in an owner complaint because the:

1. Heating capacity seems to be decreasing.

2. Heating is uneven and drafty.

3. Operating cost is rising.

4. Noise level is rising.

One adjustment that is required by the passing of time is the reduction in capacity due to dirt accumulation on the air filter. Every 5 percent reduction in air flow rate results in a capacity reduction of about 1 percent. Dirt which is allowed to accumulate on the blades of the fan, dampers which restrict air flow unnecessarily, and rugs or furniture placed so that air cannot move freely through a supply outlet or return grill will all have the same effect upon capacity as the dirty filter.

Even though collectors check out as operating properly, it is quite possible for the owner to complain of improper operation due to poor air distribution. This is a matter of damper adjustment. The dealer-contractor can help ease this problem by making branch dampers accessible and by marking the various required damper positions.

To avoid drafts, registers with adjustable blades are preferable. A system cannot always be designed with outlets located so they may serve with a single setting. For this reason, compromise outlet locations must be fitted with types of registers which have the necessary flexibility so that they may be adjusted seasonally to throw the conditioned air in a pattern which will not cause discomfort.

Increased operating cost is not always the result of a malfunctioning system. Sometimes it is due to a change in the owner's operating schedule.

Complaints about noise result from: (1) adjustments to air flow, (2) changes in damper positions, (3) wornout equipment isolators, (4) wornout pump bearings and belts, and (5) loosened fasteners. Generally, a noise which develops with system use is not as difficult to correct as an original noise.

COMMON PROBLEMS

Reports on solar demonstration projects involving both solar domestic water heating and whole house space heating included these observations:

1. Large heat losses from inadequately insulated pipes and ducts.

2. Malfunctioning controls.

3. Failure of drain-down systems to drain completely (thereby causing freezeups).

4. Noise from solenoid control valves.

5. Freezeup in inadequately protected pipes leading to closed cycle freezeable collectors.

6. Leakage in antifreeze loops.

7. High power consumption from oversized fans or inadequate ductwork.

8. Leaky valves and dampers.

9. Leaks in closed-loop circuits diluting antifreeze solutions.

10. Not taking into account different properties of heat transfer fluids when sizing expansion tanks and pumps.

11. Improper connections to storage tanks (not taking full advantage of temperature stratification inside tank).

12. Plastic covers on some collectors fluttering in high winds.

13. Incompatibility between some heat transfer fluids and seals and gaskets used.

14. Weep holes in collectors becoming plugged and causing pressure buildup behind glass covers.

15. Overheated collected causing damage to collector and nearby building materials or components.

16. Out gassing of materials inside collector and depositing a film on the inside of the collector (solid materials vaporize).

Finally, two of the most important errors contributing to solar heating problems were the inadequate calculation of building heat loss and the use of inaccurate weather data.

WARRANTIES

The types of warranties offered by manufacturers of solar heating equipment vary considerably. At the present time, if a supplier provides any warranty, it is of the "limited" type. Under its terms, the equipment is warranted to be free of defects in materials and workmanship, and that, if such defects are found within a certain period of time after initial use, correction or replacement will be made without cost to the user. Most of the suppliers of solar equipment do not currently offer any type of warranty. A few larger companies involved in solar equipment manufacture are offering one-year limited warranties. One company marketing an air system offers a 10-year limited warranty.

There appears to be no manufacturer's guaranty as to thermal efficiency or heat delivery capability of solar equipment. Although manufacturers are providing that type of information in their sales literature, they are not guaranteeing the performance in the field. To a certain degree, this omission is due to the inability of the manufacturer to control the quality of the installation. In addition, manufacturers supplying only certain components of a system, such as the collector, cannot be assured that the other components in the system are correctly selected or integrated with their own product. Thus, inferior performance might well be due to factors other than those controlled by the collector manufacturer. A performance warranty would thus be difficult to establish and maintain.

Still another problem in providing a meaningful performance warranty is the great variation in climate encountered and the practical difficulty in accurately measuring the output of the installed equipment. Instrumentation is usually not provided, so measurement of performance is likely to be an expensive investigation by an experienced engineer. Disputes, litigation, and other problems are inevitable.

Practical performance warranties should become available for complete solar heating systems provided by a single manufacturer when assembled and installed by a single responsible individual or firm. Under such conditions, the manufacturer has sufficient control of the system design and the quality of the installation to have assurance of performance. The manufacturer could then guarantee the system to the installing firm which, in turn, would guarantee it to the purchaser. In case of dispute, the installer could measure system performance in the presence of the owner and a third party, if demanded for determination of conformance. If inadequacies are determined, corrections would be made in compliance with the warranty, and the installer and manufacturer would establish responsibility for the departure from specifications.

Such developments as the Home Owners Warranty (HOW) program, sponsored by the National Association of Home Builders, can be expected to have an influence on solar heating equipment guarantees. Under the HOW program, all defects in a residential structure will be corrected at no cost to the owner during the first three years of use. It may be expected that solar heating equipment will have warranties conforming with such a program. Manufacturers will then be required to guarantee to the dealer and installer the necessary support for compliance with this program.

The solar equipment manufacturing industry unfortunately includes several small suppliers having practically no experience with solar equipment and offering no warranties of any kind. Purchasers of such equipment have very little chance of reimbursement for costly failures. Even if a small, marginal manufacturer offers some sort of warranty, a purchaser does not have much assurance that the manufacturer will remain in business long enough to make good on its guaranty. In the event of equipment defect or failure, the owner (or contractor if an installation guaranty was provided), would suffer the loss. These and other topics are discussed in the government report, *"Buying Solar,"* published June 1976 by the Federal Energy Administration and HUD.

CONSUMER INFORMATION

Many homeowners are not interested in the mechanical operation of a solar system any more than they are about the mechanical operation of a car. However, some homeowners will be interested in complete details of the system. The contractor will have to decide, perhaps from the questions asked, just how much explanation of the system is necessary.

Also, the Consumer Product Safety Act of 1972, in addition to emphasis on the design and marketing of unsafe products, also stresses that essentially hazardous products must be properly *labeled* and full and complete *instructions* provided. The Consumer Products Safety Commission has stated

"They (manufacturers, wholesalers, and dealers) must be in a position to advise the buyer competently on *how to use* and *how to maintain* and repair the product (sold)."

Here are a few suggestions.:

1. After the system has been installed, a qualified person who is familiar with the operation of the system should place the equipment under all modes of operation to insure that it is functioning properly. The owners should be shown the location of the fuse disconnect switch and the thermostat, and be instructed in how to start, operate, and stop the units and adjust temperature settings. The manufacturer's installation and operating instructions should then be *delivered* to the owners and *reviewed* with them.

2. All related component installation and service instructions should be placed in an attractive binder with your company's logo, name, address and phone number placed on the cover, together with a brief letter of welcome in the case of a new building purchase, reference to your guaranty or other similar introductory statement on quality of materials, workmanship, etc. This binder should be stored in a prominent space in the equipment room.

3. Whether or not a service contract is written, or if the equipment is under warranty, the owner should obtain the dealer's name and service department telephone number. A copy should also be permanently placed on the unit.

4. To avoid unnecessary calls and extensive interruptions in comfort heating, it is often convenient to provide a check list in case of failure. Things to do *before* calling for service might include checking for: blown fuses or open circuit breakers, dirty filters, broken fan belt, open dampers, a thermostat that is set too high, etc. Both the owner and the contractor benefit from such a simple check-out routine.

OPERATIONAL PROBLEMS

A report on the operation of various HUD demonstration programs was prepared for The National Solar Heating and Cooling Information Center by Dubin-Bloone Associates in July, 1977. That report identified the many areas of inadequacies or insufficient detail for various HUD projects. From an analysis of this report, the following data are presented:

HUD Solar Heating Projects

Problem Areas	Description

System

1. building heat load calculation errors by designer and/or contractor
2. rule-of-thumb estimating is totally inadequate
3. failure to seek help from the system component manufacturers
4. lack of technical information (example: flammability to transfer fluids)
5. poor quality control and shipping practices by manufacturers (example: broken components being delivered)

Collector

1. poor packaging practices to prevent damage
2. out gassing — breakdown of material that results in a film on the inside of cover plates
3. unable to withstand 300°-400°F stagnation temperatures
4. broken tubing due to shipment in cold weather when testing water remained inside — should have been totally drained
5. weep holes plugged that allows pressure buildup in collector that results in cover plate pop-out or breakage
6. unstable selective coatings which are often experimental materials
7. moisture collection and roof rotting—prevent with above roof mounting design
8. leakage from cracks, breaks or faulty clamps in 10% of the systems evaluated

Storage

1. fewer problems with low (atmospheric) pressure tank installations
2. temperatures exceeding the 160°F rating for fiberglass tanks
3. underestimating liquid thermal expansion (1.25 multiplier for glycols)
4. misunderstanding of stratification
5. better out-of-the-way placement was feasible

Fluids

1. corrosion due to lack of inhibitor additives
2. glycol solution (50%) diluted by automatic fill which only adds water
3. corrosion caused by failure to use "getter column" if aluminum collector is installed

Piping, Pumps and Valves

1. inadequate insulation resulting in lower system efficiency or freeze-up especially in uninsulated parts of the structure and when pipes pass through the walls, floors, and ceilings
2. heat loss due to failure to insulate duct joints
3. freeze-up of lack of drain-down because piping was not pitched properly
4. poor quality material used that caused early component failure
5. gaskets dissolved by some fluids

SUMMARY

As the Dublin-Boor report suggests, the solar heating system is in its infancy. The industry is still experiencing developmental problems. Therefore it is necessary that designers, architects, engineers, manufacturers, installers and even consumers continue to keep up-to-date on the latest developments in the field. Technological advances are being made and introduced to the industry almost on a daily basis. Efficient and effective means of utilizing this natural energy source will soon be realized.

LEGAL RESPONSIBILITIES

As the supply of fossil fuels becomes more critical in the future, alternative energy sources will have to be developed. Solar energy is one of the alternatives that is gradually becoming more adaptable to heating purposes. The technology is in its infancy but steady growth is anticipated. The concern of this lesson is not with identifying the technology of solar heating but with the impact that technological development has on people and property. The lives of people are affected where their health and safety are concerned. The value of solar heated property is affected by such factors as taxes, easements, covenants, and zoning. People must be protected (1) by the installation of reliable equipment made as hazard-free as possible by the manufacturer, and (2) by being informed by the builder or contractor as to operating and maintaining procedures to be followed when using the equipment. Justification for incentives to install, and rights and privileges of ownership is continuing to be established by federal, state, and local legislative bodies.

CONSIDERATION FOR PEOPLE

There are some aspects of the solar heating system in which the consumer has invested and must be informed concerning those aspects that relate to health and safety, because certain operational and maintenance problems may be delegated to the owner or occupant of the structure. Equipment placement is important since various monitoring and maintenance procedures must be followed. For example: the air filters need to be changed regularly, liquid filters must be cleaned or replaced regularly, liquid levels as well as antifreeze concentrations must be maintained, pumps and blowers should be lubricated at certain intervals, and fluid leaks must be stopped as they occur and are located.

Therefore, the occupant should be informed about: (1) turning off electrical service before proceeding with certain service tasks (2) which components may be excessively hot and cause burns, (3) slippery conditions caused by spillage or leakage of antifreeze solutions, (4) the presence of guarding around moving parts, (5) any problems of toxicity or probable skin irritation from chemicals used in adhesives, (6) heat transfer materials and components, and (7) any other related matters.

Unfamiliar sounds may have to be explained. The occupant may be concerned about what noises are transmitted during the cycling of operational modes. Depending on the placement of equipment and the noise tolerance limit of the occupant, a decision may be necessary to add some acoustical material to the equipment area.

Water contamination could be a problem. Introducing non-corrosive additives into potable water (that which is used for cooking, bathing, drinking) must be prevented. Therefore, some labeling procedures may be good insurance for the system designers and installers to warn occupants about possible contaminants.

CONSIDERATION FOR PROPERTY

The knowledge and practices of solar heating as applied to structural heating comprise a relatively new technology. In many communities, this alternative fuel source has been recognized. In others, no efforts have been made to accommodate this form of heating system and no special considerations for rights and responsibilities concerning solar heated property have been addressed.

Right to Light

Regardless of the percentage of the heat load designed into solar heating equipment, the most critical ingredient to the function of the system is solar energy. When structures are constructed, there is an obligation to maintain solar exposure indefinitely. This means that space above and to the sides of a solar home must be free of obstructions. There is little difficulty in guarding against this problem in low-density building areas. The problems exist for high density mutliple-story areas. See Figure 11-1

Problems arise from both structural interference and vegetation shading. A tall building placed where a projected shadow is cast on the collectors greatly reduces the efficiency of the solar heating system. The same problem exists where trees grow to tall and shade the collectors. There must be some legal procedures for zoning and

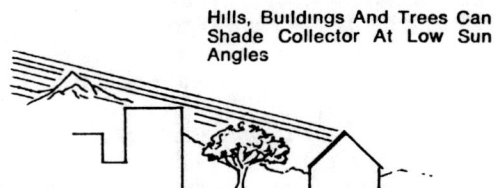

Hills, Buildings And Trees Can Shade Collector At Low Sun Angles

Fig. 11-1. Right to light may pose new problems with neighbors.

topping the vegetation even when it is not on the same property as the solar home. Some states (e.g., Florida, Colorado, and New Mexico) have proposed legislation to prevent neighbors from planting trees and shrubs that could interfere with collector insolation. Similar measures are being considered in other parts of the United States as well as Australia, Brazil, Canada and Sweden.

Freedom From Glare

Not only must the occupant of solar heated property have rights to light, but protection of others from glare is another issue. Neighbors and passerbys, either walking or riding in a vehicle, should *not* be exposed to glare from a solar collector. This problem may be handled by a less-than-latitude tilt which reduces collector efficiency, a solution which could interfere with the need for heat in the winter.

Zoning

One way to prevent collector inefficiency could be the development of appropriate zoning regulations. Placement of buildings on the lot must be done properly to minimize any nuisances caused by light reflection on adjacent property. Landscaping and street orientation, when laying out areas for new construction, may require some different considerations. Problems will also have to be resolved regarding building height so that shadows do not effect the function of nearby collectors especially when the solar heated structure is already in use.

Building Codes

The intent of building codes is to provide consistant guidelines for construction. In most communities, building codes do not address the problems of solar heating installation as such. However, they do provide direction for the use of conventional heating components, for example, pumps, blowers, and air ducts. Permits to install solar heating systems may have to be obtained based on vague interpretations of existing regulations. Some of the problems to be addressed are: design characteristics, intent at the time that codes were written, specific wording, attitudes of local authorities, and costs of providing criteria for an alternative testing procedures. Many of these regulations deal with structural soundness to withstand loads and resistance to natural phenomena such as hurricanes, earthquakes, or violent winds. Alternative roofing, plumbing, electrical, heating, ventilating, glazing materials, and work practices

may have to be negotiated with local authorities before permits can be issued and construction or retrofitting started.

As governmental agencies update their codes, problems of solar heating systems will have to be included. Some of the areas of concern will include: health and safety of the occupants; fire protection; potable water supplies; pressures within collectors, tanks, and other solar components; temperature ranges particularly as they affect thermal expansion; transfer mediums, window and door glass; roof loads; mechanical devices; and non-traditional construction methods and materials.

The SMACNA publication, *Installation Standards,* included with this training package is a good example of an useful installation guide for planning and installing solar systems for one and two family dwellings.

Another document, *HUD Intermediate Minimum Property Standards Supplement 1977 Edition, Solar Heating and Domestic Hot Water Systems,* presents minimum dwelling standards for (1) one family (2) two family, (3) multi-family, and (4) care-type (nursing) facilities. The publication is too comprehensive for summarization in this course. However, it is an extremely valuable reference book for designers, installers, and servicepersons. Chapters included in the HUD Supplement deal with (1) general use of the document, (2) site design, (3) building design, (4) product and construction materials, and (5) construction practices. References in the appendices relate to (1) calculation procedures for hot water and space heaters thermal performances; (2) standards for various materials; (3) system and component illustrations; (4) definition of terminology; (5) reference standards; and (6) identification of abbreviations for various code groups, associations, and governmental agencies involved in the publication of the document.

One very comprehensive effort to deal with coding problems resulted in the publication *Uniform Solar Energy Code (USEC)* by the International Association of Plumbing and Mechanical Officials (IAPMO). The purpose of the document is to provide minimum standards and requirements for the protection of the public health, safety, and welfare. Architects and engineers can use this publication when preparing drawings and specifications for solar heating installations. It deals with the administration of the code, definition of terms, and also addresses other specific issues. Details from the various chapters of the *USEC* are summarized in the following section of this lesson.

Quality and Weights-Alternate Materials-Alternate Methods of Construction

This particular chapter in the IAMPO publication presents a brief description of standards that have been prepared through the cooperation of various organizations.

GENERAL REGULATIONS—GENERAL INSTRUCTIONS AND REQUIREMENTS

This section deals with (1) the requirement for plans; (2) improper location of equipment; (3) prohibited fittings and practices (use of dissimilar metals); (4) retrofitting; (5) protection of piping, materials, and structures; (6) hangers and supports for horizontal and vertical piping; (7) trenching, excavating, and backfill; (8) changes in direction (appropriate use of fittings); (9) inspecting and testing; (10) maintenance; (11) existing construction (only if a building is dangerous and a menace to life, health, and property must it be altered when installing the solar system); (12) health and safety; (13) abandonment; (14) other systems (DHW, swimming pools, etc.); and (15) detailed requirements (e.g., safety requirements, controls, and welding).

Piping. Such topics as: (1) unlawful connections; (2) cross-connection control (requirement for devices to prevent back-flow when common potable water, non-potable water, and/or sewer connections may occur); (3) materials (identifies approved materials); (4) valves (sizing and placement); (5) water pressure, pressure regulators, and pressure relief valves (use and capacity); and (6) automatic air vents are reviewed.

Joints and Connections. In this chapter, requirements for the following practices are identified: (1) tightness (gas and watertight); (2) types of joints (threaded, soldered, flared, and compression fittings); and (3) special joints (tube to threaded pipe, brazing or welding, expansion joints, unions, ground joints, flared, or ferrule connections).

Tanks. Specifications for construction and use of tanks are in the section. They deal with: (1) storage or heat exchanger tank construction (general specifications related to use of concrete and steel, and (2) expansion tanks (open or closed systems and minimum capacity for closed system).

Collectors. Only the general construction specifications and location requirements are described in the chapter. (They have been presented in Lesson Three.)

Insulation. Specifications for insulation as they relate to general requirements for pipes, ducts, and tanks are discussed in the section.

Ductwork. The chapter merely recognizes that duct installation is a part of the solar heating system. It states that the Mechanical Code is the regulatory reference for sizing and locating ductwork.

This has been only a brief summary of the *Uniform Solar Energy Code.* The major purpose for including it is to provide installing and servicing technicians with the areas of concern identified by IAPMO for efficient and reliable solar heating systems.

As a general rule, an owner or contractor planning to install a solar heating system should contact the local building inspector *prior* to the expenditure of major effort on the project in order that any questions which may relate to compliance with the code could be resolved in advance. If a particular solar heating system or component clearly violates a code requirement, a change to some other type of hardware could be made prior to expenditure of significant funds on a system which would not be acceptable.

Components that comprise a solar heating system must meet acceptable standards of quality. These standards are established by extensive research and development activities. Various organizations are involved in developing specifications for the quality of products that are used. Material standards for solar heating components are created by the American National Standards Institute, American Society for Testing and Materials, International Association of Plumbing and Mechanical Officials, Underwriters' Laboratories, SMACNA, ASHRAE, and the federal government which contracts with other R&D organizations and then publishes the results of their decisions.

Installation

Along with quality components, there is also a need for competent mechanics to install the equipment. The need for qualified electricians to install electrical and electronic controls generally is not questioned. However, jurisdiction over the erection and rigging of collectors has become a negotiable problem.

The problem of who should install collectors has recently been raised in negotiating construction contracts. Two groups have been involved. They are the United Association of Journeyman and Apprentices of the Plumbing and Pipe Fitting Industry of the U.S. and Canada (USJAPPFI) and the Sheet Metal Worker's International Association (SWMIA).

The jurisdiction agreement which these two organizations have reached is that the supporting and rigging of solar collectors with tubing and/or piping for liquid fluids will be installed by a composite crew equal in size to the numbers of members of the respective unions. For air fluid systems, the collector array as well as all ducts throughout the fluid transfer system, will be installed by SMWIA. For the liquid fluid systems, all pipes and valves will be installed by UAJAPPFI.

Appraisal

Appraisal practices for financing a solar heating system during the planning phases is difficult. The market has not been developed sufficiently to provide guidelines for establishing a market value. Appraisers may choose to ignore the system, treat it as a hot water system, or limit their assessment to the installation cost. On the other hand, the entire system may be considered a liability by some individual appraisers.

Appraising for tax purposes is another matter. This is especially true if the construction loan appraiser chose to ignore the system. Then the borrower pays separately for a heating system that has been ignored but then pays taxes based on the total pacckage—structure plus system.

Equitable solutions to problems of financing and taxing solar heated structures are being discussed. Many federal, state, and local legislative bodies have introduced and/or passed acts which relate to problems of construction loans and taxation as incentives to invest in solar heating systems.

Incentives

Legislation to provide incentives for investments in solar heating systems could be considered regulatory or compensatory. Legislated regulations deal with such matters as establishing solar energy policy agencies that would respond to problems such as right to light, freedom from glare, nuisance matters, and building standards or codes.

Compensatory legislation would deal with low cost construction loans and tax breaks to those who invest in solar heating systems. Low cost loans are being proposed for both private citizens and industries. Some considerations are being given to investment tax credits and amortization procedures. Tax reducing factors are being proposed for income, sales, and real estate. Federal, state, county, or city income and/or property taxes may be reduced if a large amount of money is invested in an entire solar heating system or one merely insulates a structure. Sales and use taxes may be discounted or eliminated on solar heating systems. Lowering the real estate

tax for a given number of years is being considered if the owner invests in solar heating when building or retrofitting a structure.

Some tax incentives that have been assessed by ERDA and presented for consideration are:

1. Purchaser tax credit of 40% on first $1,000 of system cost, 25% on next $6,400 of system cost, with up to a maximum of $2,000 tax credit, effective 1978-1982.

2. Installer investment tax credit of 20%, effective 1978-1982. (Because this incentive showed little promise early in the calculations, it was omitted from detailed consideration.)

3. Builder/developer tax credit of 20% of total solar heating and air conditioning system (SHAC) cost or Incentive I, effective 1978-1982.

4. Loans of 5% for 20 years for residential retrofit applications, effective 1978-1985.

5. Purchaser tax credit of 25% of the first $2,000 and 15% of the next $6,000 of system cost up to a maximum of $1,400 tax credit, effective 1978-1982.

6. Purchaser tax credit of 40% of the first $1,000 of system cost and 25% of the next $6,400 of system cost for 1978 and 1979, 30% of the first $1,000 and 20% of the next $6,400 for 1980 and 1981, and 25% of the first $1,000 and 15% of the next $6,400 for 1982 and 1983.

Insurance

Insurance companies have not been reluctant to provide insurance for solar heated structures. Comprehensive insurance has not yet become a high risk factor since glass breakage has not been excessive. Fire insurance should not be affected since the required additional insulation is fire resistant and the energy requirements (electricity for controls, etc.) are similar to those used in conventional heating components. There may be a need to have fire stops placed in the roof-to-equipment-room space where pipes are located. These air columns could act as a flue and cause a draft that would draw flames from the basement to the attic. Also, heat and smoke detectors should be installed as recommended by Underwriter's Laboratories.

Warranties

Warranties have long been used as a form of consumer and manufacturer protection. Many have been written in terms which caused much confusion especially to those who attempt their interpretation. As a result, an effort has been made by the federal government to give directions for writing and interpreting warranties. This action has resulted in the passage of the Magnuson-Moss Warranty-Federal Trade Commission Act (FTC) of 1975.

There ared three types of warranties: *full*, *limited*, and *implied*. A *full* warranty as now regulated by the FTC stipulates that:

1. The product must be repaired at no cost to customer.

2. The product must be repaired within a reasonable amount of time.

3. The purchaser is required only to satisfy the manufacturer that service is requested except if the warrantor can establish that there is cause to suspect the proper use of the product by the purchaser.

4 The warranty applies to anyone who owns the product during the duration of the warranty.

5. If the product is not repaired within a reasonable time, the manufacturer must offer a replacement or refund. If cash is paid to the customer, the product must be returned to the manufacturer.

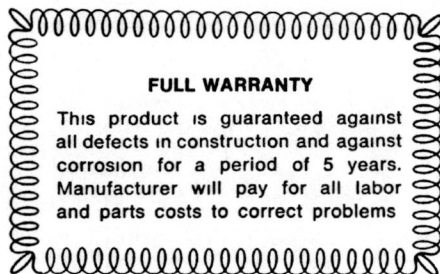

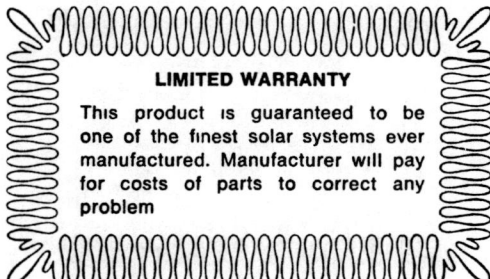

This full warranty may cover the entire product or only parts of it. It will also contain the duration of time that the warranty is in effect.

A *limited* warranty:

1. Covers parts and none or only a part of the labor expenses.

2. Arranges pro-rate refund or credit to allow purchaser to get some percentage of the purchase price refunded.

3. States that the purchaser must pay for and be responsible for the return of the product.

4. Provides that only the original purchaser is protected.

Within the framework of limited warranties, manufacturers are required to identify the warranty duration period, limitations, and exclusions.

An *implied* warranty is the result of state legislation to protect the customer. The most common implied warranty is that the product will perform the ordinary purpose for which it was designed. When a manufacturer does not prepare a written warranty or is not explicit about its content, the implied warranty provides legal recourse on the part of the purchaser.

Warranties must accompany the product and be prominently displayed on or near the product and/or its packaging. In the case of a solar assisted heating system, different warranties may cover different components. Therefore, the purchaser should be given a packet containing the warranties when the system has been installed.

REACTION

A lesson about the legal responsibilities concerning people and property should be open-ended and never summarized. The technology of solar heating is in a state of evolution. Information regarding new and improved materials and processing create different problems requiring negotiable resolutions. These problems have to be dealt with individually as they affect the lives of people, their property, and the environment.

FULL WARRANTY

This product is guaranteed against all defects in construction and against corrosion for a period of 5 years. Manufacturer will pay for all labor and parts costs to correct problems

LIMITED WARRANTY

This product is guaranteed to be one of the finest solar systems ever manufactured. Manufacturer will pay for costs of parts to correct any problem

Fig. 11-2. Plain language warranty information prominently displayed is now covered by Federal law.

SELECTED REFERENCES

Installation and Maintenance of Solar Heating Systems; College of Engineering, Institute of Energy Conversion, University of Delaware, Newark, Delaware.

Solar Heating and Cooling of Residential Buildings, Sizing, Installation and Operation of Systems, Solar Energy Applications Laboratory, Colorado State University, Fort Collins, Colorado. Also available from U.S. Government Printing Office; Stock No. 003-011-0085-2.

Solar Heating Systems Design Manual; Training and Education Department, Fluid Handling Division, ITT Bell & Gossett, Morton Grove, Illinois.

Intermediate Minimum Property Standards Supplement, Solar Heating and Domestic Hot Water Systems, U.S. Department of Housing and Urban Development, Washington, D.C.

Applications of Solar Energy for Heating and Cooling Buildings. ASHRAEE GRP 170, American Society of Heating, Refrigerating and Air Conditioning Engineers, Inc., New York, New York.

Solar Heating of Buildings and Domestic Hot Water, U.S. Department of Commerce, National Technical Information Service AD-A021 862; Springfield, Virginia.

Application Guide for Contractors, Builders, Solar Assisted Heat Pump; Fedders Corporation, Edison, New Jersey.

Buying Solar, Federal Energy Administration; Superintendent of Documents, U.S. Government Printing Office, Stock No. 041-018-00120-4, Washington, D.C.

Solar Water & Space Heating—An Economic Analysis, Division of Solar Energy, ERDA, Superintendent of Documents, U.S. Government Printing Office, Stock No. 060-000-0038-7, Washington, D.C.

Maintenance Cost of Solar Air Heating Systems, Solar Energy Applications Laboratory, Colorado State University, Fort Collins, Colorado

ASHRAE Standard 93-77, *Methods of Testing to Determine The Thermal Performance of Solar Collectors,* American Society of Heating, Refrigerating and Air Conditioning Engineers, Inc. New York, New York.

Uniform Solar Code, International Association of Plumbing and Mechanical Officials.

Relative Areas Analysis of Solar Heating System Performance, Master's Thesis by C. Dennis Barley, Colorado State University, Fort Collins, Colorado.

General Services Administration, Public Building Service. *Energy Conservation Design Guidelines for New Office Buildings,* Second Edition. GSADC75-12219, July, 1975.

APPENDIX A

Climatological Characteristics for Selected Locations in the United States and Elsewhere

State and City	Latitude (°N)	Elevation, feet (E)	Jan. Temperature °F. (T)	Jan. Solar BTU (day)(ft²) (s)	(°F) January (d$_j$)	(days) Annual (d)
ALASKA						
Annette Island	55	110	32.0	236	949	7,192
Barrow	71	22	-13	13		
Bethel	61	125	6.8	136	1,903	13,196
Fairbanks	65	436	-11.2	70	2,359	14,279
Matanuska	62	180	12.2	747	1,646	10,849
ARIZONA						
Page	37	4,270	32.0	1,105	1,063	5,380
Phoenix	33	1,112	50.0	1,093	474	1,765
Tucson	32	2,556	50.0	1,152	471	1,800
Yuma	32	199	53.6	1,124	308	1,006
ARKANSAS						
Little Rock	35	265	41.0	729	756	3,219
CALIFORNIA						
Davis	39	51	44.6	581	583	2,502
Fresno	37	331	44.6	605	605	2,611
Inyokern	36	2,440	44.6	1,148	546	2,122
La Jolla				900		
Los Angeles	34	99	53.6	946	372	2,061
Pasadena	34	864	53.6	925	343	4,794
Riverside	34	1,020	51.8	1,013	406	1,874
San Diego	32	19	53.6	976	314	1,507
Santa Maria	35	238	50.0	975	459	2,967
Soda Springs	39	6,752	20.0	823		
COLORADO						
Boulder	40	5,385	32.0	740	992	5,540
Grand Junction	39	4,849	26.6	854	1,209	5,641
Grand Lake	40	8,389	15.8	781	1,556	10,802
FLORIDA						
Apalachiccola	30	35	53.7	1,078	347	1,308
Belle Isle	27	31	71.0	1,096	98	
Gainesville	30	165	55.4	1,023	332	1,239
Jacksonville	30	24	53.6	984	348	1,327

APPENDIX A

Climatological Characteristics for Selected Locations in the United States and Elsewhere

State and City	Latitude (°N)	Elevation feet (E)	Jan. Temperature °F. (T)	Jan. Solar BTU (day)(ft²) (s)	(°F) (days) January (d$_j$)	(°F) (days) Annual (d)
Key West	24	6	69.8	1,205	16	64
Miami	26	9	66.2	1,263	74	214
Pensacola	30	13	51.8	921	427	1,578
Tallahassee	30	58	51.8	909	375	1,485
Tampa	28	11	60.8	1,204	202	683
GEORGIA						
Atlanta	34	976	42.8	834	636	2,961
Griffin	33	980	42.8	876	505	2,136
HAWAII						
Honolulu	21	7	75.0	1,339	0	0
Mauna Loa	21	20	40.0	1,926		
Pearl Harbor				1,325		
IDAHO						
Boise	44	2,844	30.2	522	1,113	5,809
Pocatello			24.8	608	1,296	6,603
Twin Falls	42	4,148	28.4	600	1,159	6,324
ILLINOIS						
Chicago	41	610	24.8	353	1,265	6,639
Lemont	42	595	24.8	629	1,265	6,639
INDIANA						
Indianapolis	40	793	28.4	541	1,113	5,699
IOWA						
Ames	42	1,004	19.4	640	1,370	6,588
KANSAS						
Dodge City	38	2,592	30.2	953	1,051	5,086
Manhattan	39	1,076	30.2	117	1,122	5,232
KENTUCKY						
Lexington			32.0	629	946	4,686
Louisville	38	474	32.0	604	983	4,640
LOUISIANA						
Lake Charles	30	12	51.8	880	363	1,459
New Orleans	30	3	51.8	788	363	1,385
Shreveport	32	252	46.4	832	552	2,184

APPENDIX A

Climatological Characteristics for Selected Locations in the United States and Elsewhere

State and City	Latitude (°N)	Elevation feet (E)	Jan. Temperature °F. (T)	Jan. Solar BTU (day)(ft^2) (s)	(°F) (days) January (d$_j$)	(°F) (days) Annual (d)
MAINE						
Amherst			24.8	427	1,339	7,469
Caribou	47	628	10.4	504	1,690	9,767
Portland	44	63	23.0	578	1,339	7,131
MARYLAND						
Annapolis		65	33.8	645	946	4,548
Silverhill	39	291	35.6	670	871	4,224
MASSACHUSETTS						
Amherst	42	750	29.0	428	1,136	
Blue Hill	42	629	28.0	555		
Boston	42	29	30.2	511	1,088	5,634
Cambridge	42	80		565		
Churchill			-16.6	239	2,558	16,728
East Wareham	42	18	32.0	504		
Lynn			30.2	434	1,088	5,634
Natick			26.6	559	1,271	6,969
Pittsfied	42	1,170			1,339	7,578
MICHIGAN						
East Lansing	43	856	23.0	423	1,262	6,909
Sault Ste. Marie	46	724	15.8	489	1,525	9,048
MINNESOTA						
St. Cloud	45	1,034	10.4	625	1,702	8,876
MISSOURI						
Columbia	39	785	30.2	662	1,076	5,046
MONTANA						
Glasgow	48	2,277	10.4	567	1,711	
Great Falls	47	3,664	21.2	508	1,349	7,750
Summit	48	5,213	15.8	449	1,538	10,625
NEBRASKA						
Lincoln	41	1,189	24.8	699	1,237	5,867
N. Omaha	41	978	23.0	752	1,355	6,612

APPENDIX A
Climatological Characteristics for Selected Locations in the United States and Elsewhere

State and City	Latitude (°N)	Elevation feet (E)	Jan. Temperature °F. (T)	Jan. Solar BTU (day)(ft²) (s)	(°F) (days) January (d_j)	Annual (d)
NEVADA						
Ely	39	6,262	24.8	876	1,308	7,730
Las Vegas	36	2,162	42.8	1,027	688	2,709
Reno	39	4,404	32.0	862	1,027	6,022
NEW HAMPSHIRE						
Mt. Washington	44	6,262	9.0	432	1,809	7,866
NEW MEXICO						
Albuquerque	35	5,314	34.0	1,134	930	4,348
NEW JERSEY						
Seabrook	40	100	40.0	592		
Trenton	40	144	32.6	637	1,020	4,947
NEW YORK						
Albany			23.0	456	1,311	6,875
Ithaca	42	950	23.0	449	1,271	6,624
New York	41	52	32.0	537	973	4,811
Sayville	41	20	35.0	603		
Schenectady	43	217	23.0	478	1,283	6,650
Upton	41	75	35	583		
NORTH CAROLINA						
Cape Hatteras	35	7	46.4	898	580	2,612
Greensboro	36	891	37.4	754	784	3,805
Hatteras	35	7	50	892	580	2,612
Raleigh	35	433	41.0	876	725	3,393
NORTH DAKOTA						
Bismark	47	1,660	8.6	581	1,708	8,851
OHIO						
Cleveland	41	805	28.4	456	1,159	6,351
Columbus	40	833	30.2	475	1,088	5,660
Put-In-Bay	42	575	26.6	441	1,159	7,211
OKLAHOMA						
Oklahoma City	35	1,304	35.6	939	868	3,725
Stillwater	36	910	35.6	762	893	3,860

APPENDIX A

Climatological Characteristics for Selected Locations in the United States and Elsewhere

State and City	Latitude (°N)	Elevation feet (E)	Jan. Temperature °F. (T)	Jan. Solar BTU (day)(ft²) (s)	(°F) January (d$_j$)	(days) Annual (d)
OREGON						
Astoria	46	8	41.0	338	753	5,186
Corvallis	44	221	37.4	371	803	4,408
Medford	42	1,329	37.4	434	918	5,008
PENNSYLVANIA						
Philadelphia	30	7	32.0	645	1,014	4,565
Pittsburgh	40	749	28.4	346	1,119	5,087
State College	41	1,175	28.4	511	1,122	5,934
RHODE ISLAND						
Newport	41	60	28.4	570	1,020	5,804
SOUTH CAROLINA						
Charleston	33	46	50.0	931	487	2,033
SOUTH DAKOTA						
Rapid City	44	3,218	23.0	684	1,333	7,345
TEXAS						
Big Spring			42.8	986	651	2,591
Brownsville	26	20	59.0	1,056	205	600
Corpus Christi	27	43	55.4	965	304	930
Dallas	32	481	44.6	851	608	2,030
El Paso	32	3,916	44.6	1,218	685	2,700
Fort Worth	33	544	44.6	927	614	2,405
Midland	32	2,854	42.8	1,034	651	2,601
San Antonio	30	794	51.8	1,020	428	1,546
TENNESSEE						
Nashville	36	605	37.4	600	828	3,696
Oak Ridge	36	905	37.4	611	778	3,817
UTAH						
Flaming Gorge				878		
Salt Lake City	41	4,227	28.4	648	1,172	6,052
VIRGINIA						
Norfolk	36	26	39.2	766	760	3,488
Mt. Weather	39	1,725	30.2	634	1,107	5,668
WASHINGTON, D.C.	39	64	35.6	585	871	4,224

APPENDIX A
Climatological Characteristics for Selected Locations in the United States and Elsewhere

State and City	Latitude (°N)	Elevation feet (E)	Jan. Temperature °F. (T)	Jan. Solar BTU (day)(ft²) (s)	(°F) (days) January (d_j)	(°F) (days) Annual (d)
WASHINGTON						
North Head	46	194	40			
Friday Harbor	49	100		321		
Prosser	46	675	30.2	431	1,123	4,597
Pullman	47	2,550	28.4	452	986	3,995
Richland			32.0	316	1,163	5,941
Seattle	47	386	41.0	287	738	4,424
Spokane	48	1,968	28.4	434	1,231	6,555
University of Washington				247		
WISCONSIN						
Madison	43	866	17.6	564	1,473	7,863
WYOMING						
Lander	43	5,370	19.4	846	1,417	7,870
Laramie	41	7,266	21.2	824	1,212	7,381
CANADA						
Edmonton, Alberta	54	2,219	6.8	327	1,810	10,268
Kapuskasing, Ontario			-0.4	405	2,037	11,512
Montreal, Quebec			15.8	405	1,566	8,203
Ottawa, Ontario	45	339	14.0	530	1,624	8,735
Toronto, Ontario	44	379	26.6	445	1,233	6,827
Vancouver, British Columbia			35.6	279	862	5,379
Winnipeg, Manitoba	50	786	1.4	482	2,008	10,679
ISLAND STATIONS						
Canton Island				2,179		
San Juan, P. R.				1,491		
Swan Island				1,631		
Wake Island				1,616		
OTHERS						
Albrook, A. B. Panama				1,446		
Taipei, Taiwan				686		

DUCT DESIGN

In most small applications, the fan is an integral part of the manufacturer's unitary equipment. Thus, once the equipment has been selected, the fan capacity at various static pressures will be known and the duct system must be sized to match the fan performance.

In all-air systems, the total system CFM (ft^3/min) is usually determined by the installed collector area and the manufacturer's recommended flow rate, which is usually stated in CFM per square foot of collector area. For example, flow rates generally range from 2 to 3 CFM per square foot. If 400 square feet of collector is installed, the total system CFM might range from 800 to 1200 CFM (2 x 400 or 3 x 400).

In single fan arrangements, this collector flow rate must be compatible with the auxiliary heating system. Thus if gas or oil fired furnaces are used as auxiliary heat, the collector CFM should provide the recommended temperature rise through the furnace. If electric heaters are used, then the collector CFM should equal or exceed the minimum flow specified for the heaters. If mechanical cooling or a heat pump is part of the system, then the collector CFM must be compatible with flow requirements that range from 300 to 420 CFM per ton. If air flow is not compatible, then multi-speed fans might be employed.

All-air systems may also be designed using two-fans—one for the collector loop and another for the conventional backup heating system.

When collector loop and conventional air flow rates are not the same, a bypass duct, as shown in Figure B-1, can be used to balance the flow.

Sizing standards are listed in the *SMACNA Installation Standards* that are a part of your training materials. Please refer to Section 3 in these *Standards* for design criteria. For specific sizing procedures refer to the *SMACNA Duct Sizing Guide, NESCA's Manuals K* and *Q* or the *ASHRAE Handbook.*

The problem with so many practical duct sizing methods is that they have been so simplified, streamlined, and practicalized that fundamentals are very often obscured. "Cook book" methods also frequently destroy the feel of the design, and the user is seldom fully aware of all the assumptions made and qualifications built into the short-cut tables and charts intended to save time.

Going back to fundamentals can, for most people, help obtain a better prespective of the design problem. In this course then, let's review the three basic design methods—*velocity reduction, equal friction,* and *static regain,* noting their advantages and disadvantages. But first, let's really get fundamental and consider several essential relationships.

Flow Rate and Pressure Affect Fan Hp

A duct system has to be sized first to fit the space available, minimize air noise (by controlling air velocity), and move air with a *minimum* of fan

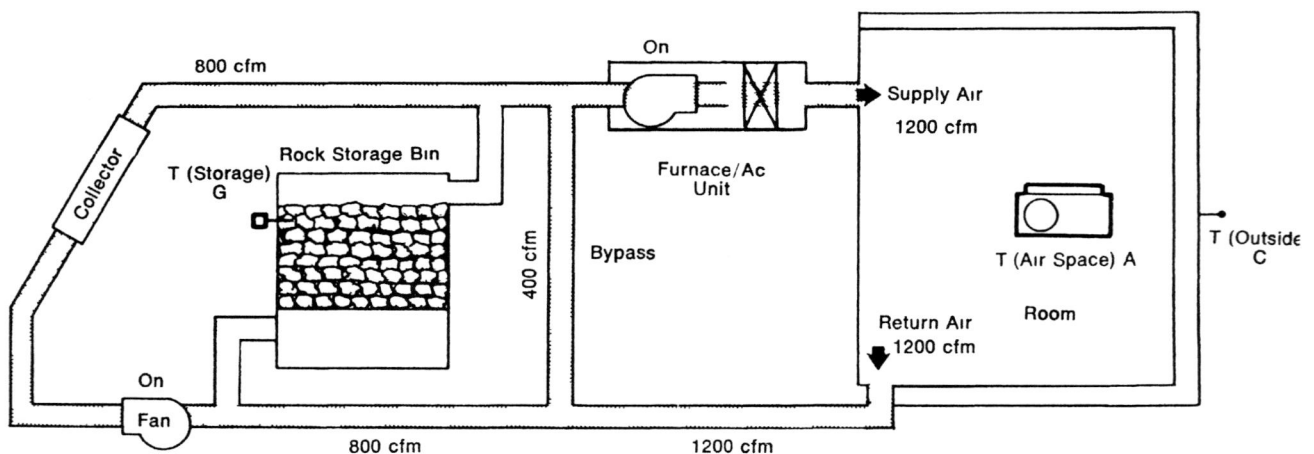

Fig. B-1.

power requirements. This last factor is often overlooked.

The horsepower needed to move air at standard conditions is equal to the product of the system air flow rate times the total pressure difference created in the system. More precisely:

$$ahp = 0.000157 \, (Q) \, (TPd)$$

where

ahp = air horsepower (divided by fan efficiency this yields brake horsepower at fan shaft)

Q = system air flow rate in cfm

TPd = total pressure difference between supply and return system, inches of water

From the formula, an increase in system flow rate (Q), brought about by an increase in heating and cooling loads, duct heat loss, or heat gains, increases fan hp and operating costs. An increase in TPd brought about by higher duct resistance on the supply or return side also increases fan hp requirements.

Total pressure (TP), in general, is the sum of the static (SP) and velocity (VP) pressures at any point in the system. The TP at the fan on the supply (or return) side is the sum of all the losses of the fittings, lengths of duct, dampers, etc. starting at the grille in the longest run and leading back to the fan. The loss through any other run will by necessity be the same. (Since all branches terminate at atmospheric pressure and are connected in parallel to the same fan, the drop in all runs is from fan pressure to atmospheric pressure.)

While TP is perhaps more desirable to work with since its value always decreases in the direction of flow, SP is in fact normally used as a basis for design. SP can increase or decrease in the direction of flow, and in tracing SP changes from grille to fan to establish system pressure at the fan, the designer must be cognizant of SP regain as well as SP loss (More on this later.)

Friction Chart Relates the Variables

One of the basic working tools in all design methods is the standard friction chart. (An excerpt is shown in Figure B-2.) This chart was developed back in 1945 based on research conducted at the old ASHRAE Laboratory in Cleveland, Ohio in which ductwork was actually subject to test.

The chart relates the friction loss or SP drop per 100 feet of straight duct (horizontal scale), to

cfm flowing (vertical scale), to round duct sizes (slant line moving up from left to right) and finally to duct velocity (slant line moving up from right to left).

Values obtained strictly speaking, apply to air between 50° and 90°F for smooth galvanized duct. Correction factors for other temperature

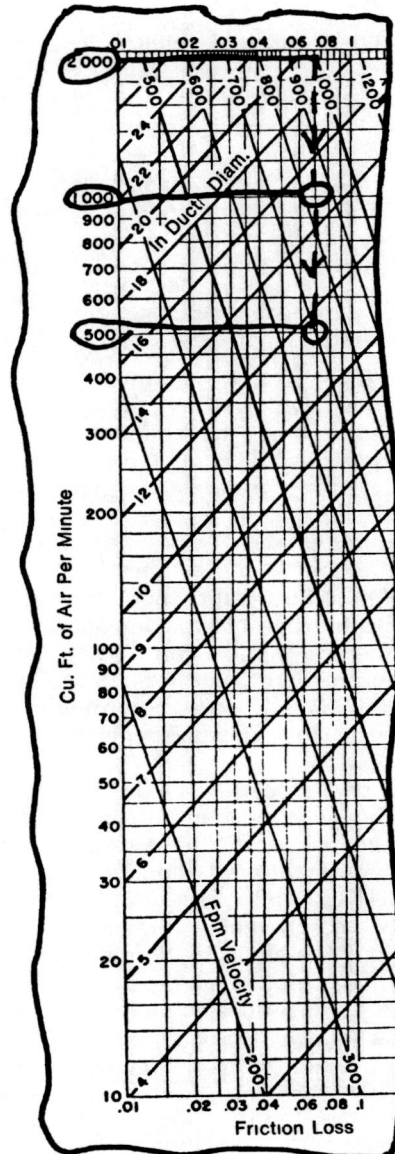

Fig. B-2. Standard friction chart relates all the variables for flow in straight ducts — pressure drop (horizontal scale), flow rate (vertical scale), duct size (slant line moving up l. to r.) and air velocity (slant line up from r. to l.).

conditions are available. Precise corrections for lined metal or for glass fiber duct have not been rigorously established as yet. (Consult manufacturers' literature.)

It's important to note that the horizontal scale indicating SP loss also indicates TP loss since in a constant size duct, velocity, or velocity pressure, is constant.

Besides the loss in straight duct, the losses caused by transition fittings, elbows, takeoffs, dampers, filters, coils, and grilles must also be included. Losses for transition fittings and elbows are usually given in terms of equivalent feet of straight duct and added to the actual duct lengths. Takeoff fittings are more complex and charts or tables list losses based on percent of flow diverted. Losses for dampers, grilles, etc. are normally provided by the manufacturer.

The losses of fittings is seen to depend on duct sizes, and as mentioned at the outset, there are three methods to size ductwork.

The *velocity reduction* method is almost no method, for it relys heavily on the designer's experience. Thus we have the paradox that the easiest method is really the most difficult method for the inexperienced designer. The method is further restricted to relatively uncomplicated layouts.

The method is merely one of assigning duct velocities throughout the system in decreasing order from fan to furtherst branch. The duct sizes needed to achieve these assigned values are selected from the friction chart. The chart also provides the pressure drop in each run of duct. The method is fast and noise control is relatively assured since the designer picks his own velocities.

The second procedure is the *equal friction* method. In this case the idea is to establish a constant pressure loss factor (pressure drop per foot) and use it throughout the system. If the system is symmetical—all runs essentially the same length—the method makes the system almost self-balancing. The equal friction method is also the most popular procedure.

To size a trunk-duct for instance, the procedure is as follows: Consider that 2000 cfm is delivered by a fan into a main. First of all, an initial velocity must be selected. With the aid of the standard friction chart, this establishes the design pressure drop per 100 ft. If we choose 1000 fpm, the pressure drop is 0.07 in. WG/100 ft. (see Figure B-2.)
Also, the main size at the fan will be approximately a 19 inch round.

Now as succeeding volumes of air are diverted into the branch from the main, it is only necessary to enter at the top of the friction chart at a loss of 0.07 and move down vertically until the horizontal line corresponding to the cfm remaining in the main is reached to determine the new main size that maintains the same loss per foot.

Thus if 1000 cfm were diverted out the first takeoff—leaving a 1000 cfm in the main—the duct size after the takeoff would be a 15 inch round. (Intersection of 1000 cfm and 0.07 in. WG in Figure B-2). If after the second takeoff, 500 cfm remained, then moving further down along the same 0.07 in. WG loss line, the main after the second takeoff should be reduced to a 11.5 inch round. This procedure is followed until the main is completely sized.

When available fan SP is known at the outset, say where unitary equipment is used, designers sometimes use this figure divided by the length of the longest run on the plans (including something for equivalent length of fittings) as a trial pressure loss factor.

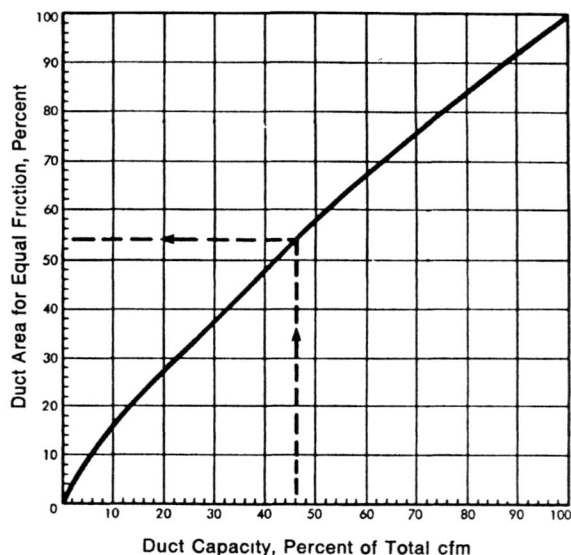

Fig. B-3. Equal friction duct sizing method is one of three basic design procedures. A constant pressure loss factor per foot of duct is established and the system is sized to provide this constant loss. Curve at left relates cfm to duct area for equal friction loss. Once an initial duct area has been determined, say by assuming an initial supply velocity, then main trunk-duct reductions and branches can be sized based on the percent of the total cfm that is carried by each run. If after a branch takeoff, the trunk carries only 46 percent of the total air supply (follow arrows) then the main should be reduced to 54 percent of the initial area.

Percent Cfm Determines Percent Duct Area

A variation of this method can be developed from fundamental flow equations that show that the ratio of duct areas is approxiamtely equal to the ratio of air flow rates in each duct area raised to the 0.8 power for equal pressure drop per foot of duct. Thus:

$$A/A_f = (Q/Q_f)^{0.8}$$

where A_f is the duct area at the fan and Q_f is the system flow rate. A and Q represent duct areas and flow rates at any other point in the system.

Figure B-3 is a plot of this relationship. Knowing the initial duct area, the duct area for any part of a main or branch can be determined from the chart.

To start the design, it's again necessary to assume a duct velocity at the fan. From Figure B-2 or the fact that cfm divided by velocity yields area, the initial duct area A_f can be determined. Let's say 4000 cfm is flowing; for a 1400 fpm initial velocity, 4000/1400 or 2.85 square foot of duct area is required. Now if 46 percent of the total air flow is diverted at the first branch, then from Figure B-3 entering the base at 46 percent and following the arrows, the branch area must be 54 percent of the initial duct area for equal friction loss/ft. Hence the branch duct area must be 0.54 times 2.85 or 1.48 square foot. Successive branches and mains are sized in a similar manner based solely on the percent of the total air volume carried.

Once sized, the actual losses through the system should be computed and compared to available fan SP, using the friction chart and available fitting loss data. If the loss is greater than the available, assume a new, larger initial duct area and repeat the procedure.

Regain Important in Big Systems

But even the equal friction method has its drawbacks. For one thing, having the same pressure drop per foot for both long and very short runs causes unbalance—similarly for the case of runs of equal length but with one having many turns and fittings. One solution is to use different friction loss rates. Branches near the fan, for instance, where system SP is still high could be sized based on a higher drop per foot than branches near the end of the main.

Another drawback in the normal use of the equal friction method is that any static pressure regain is typically ignored. While this is of little consequence in small, low velocity systems (ignoring regain merely provides a small safety factor) in larger, high pressure systems, ignoring regain can result in higher than necessary fans

sizes and power requirements. Thus, a third design procedure exists—the static regain method.

Whenever some air is turned from the main into a branch, there is a decrease in the velocity of air remaining in the main trunk. In theory, were there no friction, the decrease in velocity pressure would be exactly offset by a rise in static pressure, since SP plus VP equals TP, and TP would be a constant value throughout the system.

In reality there *is* friction, so regain is not 100 percent and the rise in SP may be only half, three-quarters or ninety percent of the decrease in VP. In a well designed system, 75 percent regain is considered likely, thus

$$SP_{rise} = 0.75 (VP_a - VP_b)$$

where VP_a and VP_b are the respective velocity pressures upstream and downstream of the takeoff junction.

Since the size of the downstream duct affects the downstream velocity—hence VP_b—the value of the static regain can be increased or decreased depending on the selection of the downstream duct diameter.

Figure B-4 shows a somewhat idealized plot of SP vs distance along a duct before and after a side-mounted register has discharged some air from the main. In this illustration the SP regain (dashed line) is exactly equal to the SP loss through the next duct section. The SP behind both registers A and B is therefore identical. In a long run of duct, where registers or branches of equal size and capacity are required it is advantageous to system balance if the same SP is made available at each junction. And in conventional SP regain design, the SP gain is in fact made equal to the SP loss of the following duct section. This is accomplished by carefully sizing the reduced section that follows.

As an example, if 4000 cfm is flowing ahead of register A in Figure B-4, from the standard friction chart a 23 inch round duct would be used upstream of A. At A, 2000 cfm is discharged out the register so that 2000 cfm continues into the next duct section. If that new duct section is sized so that there's a 1000 fpm velocity (say by using a 19.5 inch round) then the static regain would be

$$SP_{rise} = 0.75 \left[\frac{1400}{4005}^2 \quad \frac{1000}{4005} \right]^2$$

or $SP_{rise} = 0.045$ in. WG

Now if the equivalent length between A and B is 100 feet (actual length plus equivalent length of fittings), then from the standard friction chart we would find that the loss per 100 feet for a 19.5 inch round with 2000 cfm flowing is 0.07 inch WG—which is greater than the computed value of the static regain of 0.045.

Repeating the procedure, only this time assuming a lower duct velocity in section A-B, say 900 fpm, then the static regain would be

$$SP_{rise} = \left[0.75 \left(\frac{1400}{4005}\right)^2 \left(\frac{900}{4005}\right)^2\right]$$

or $SP_{rise} = 0.052$ in. WG

For a 900 fpm velocity, a 20 inch round duct is

required, and again from the friction chart the loss per 100 feet is now 0.05 or approximately equal to the regain. A 20 inch duct is therefore the desired reduction in the main. To speed up the design and avoid the trial and error approach illustrated here pre-calculated sizing tables are available.

For large systems then, a combination of static regain and equal friction sizing methods might be employed—static regain to size the mains and equal friction in the branch runs.

The intention here has been to present the fundamental concepts behind the three common design methods and to acquaint the student with the three general design objectives—size to fit space available, control noise by regulating air velocity, and minimize fan power requirements.

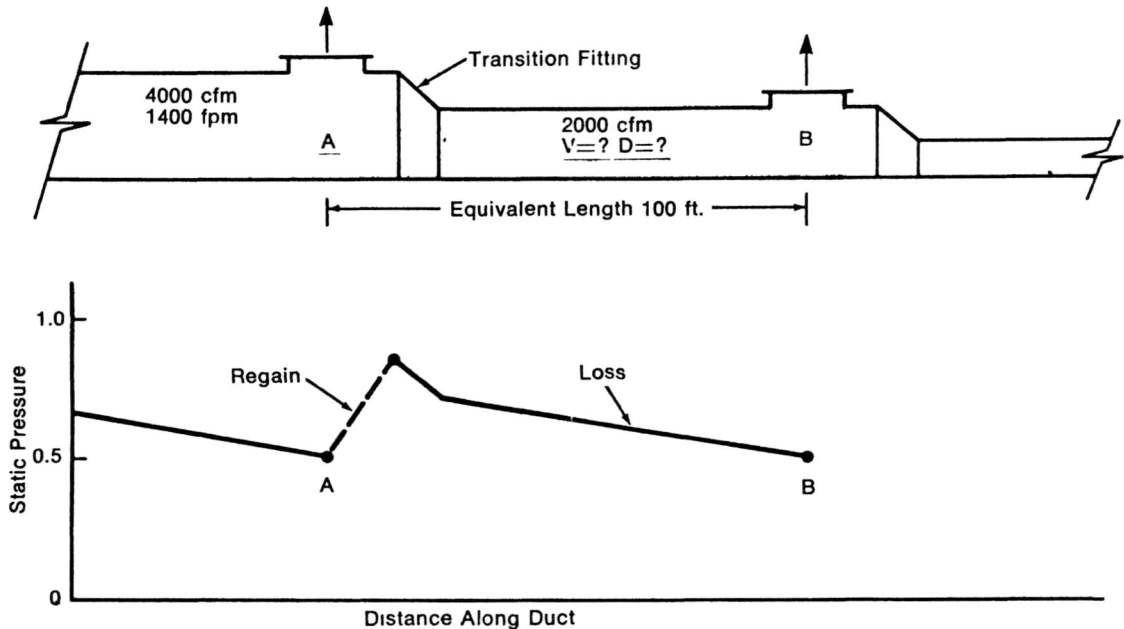

Fig. B-4. Static pressure can increase (dashed line) as well as decrease along a duct system. Static regain sizing method equates the value of the regain at A to the l[o] the following duct section (A to B).

AIR HANDLING SYSTEMS

A good air handling system is the result of a careful study of factors relating to the placement of the heating equipment, the supply outlets, and the return air grilles.

The first step in any design is obviously to study the building plans very carefully and be certain that the plans accurately describe the actual building—whether it is new or old. The designer should also confer with the builder or owner, if possible, to learn of any special needs or demands of the conditioning system.

Before planning the duct system, the best locations for the equipment and the supply outlets must be determined. A central location helps to keep variations in delivered air temperature to a minimum, and simplifies the air balancing problem.

Locating Heating Outlets

When speaking of the performance characteristics of supply outlets (Figure C-1), reference to the several effects they have upon the stream of air passing through them from the duct into the room is made. One such characteristic is called *throw*. With a knowledge of *throw*, selection of an outlet which can deliver the volume of air required without causing a high velocity jet stream to strike a wall and deflect down into the occupied space can be accomplished. Conversely, an air flow pattern which will provide adequate coverage for a space and not leave stagnant pockets or stratified layers of air to cause discomfort can thus be made

Other characteristics include air stream drop at the throw distance (distance at which air stream velocity falls to 50 fpm), *spread, sound level,* and *air pressure drop.* With the variety of air distributing devices available today and with a working knowledge of their characteristics, air can be successfully introduced into a room from any point desired.

The choosing of supply outlet locations for a year-round residential system is not a haphazard procedure. Rather, the choice is dictated by several factors which must be considered simultaneously. These might be stated as questions:

1. Will the location chosen serve for both heating and cooling with minimum adjustment?

2. Will the location allow mixing of conditioned and room air to take place in a way that will not disturb room occupants?

3. Will the location result in uniform room conditions and not short cycling of the air to an adjacent return grille?

4. Is the location likely to prove impractical relative to drapes or furniture placement?

5. Can the location be served by a branch duct of reasonable length as compared with other branches in the system?

Performance Guidelines

Generally speaking, the upward projection of conditioned air along an outside wall is the best distribution pattern for a year-round system, al-

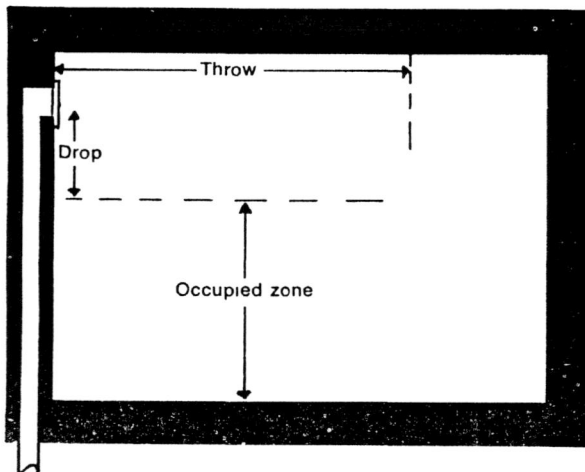

Fig. C-2. Typical air stream pattern.

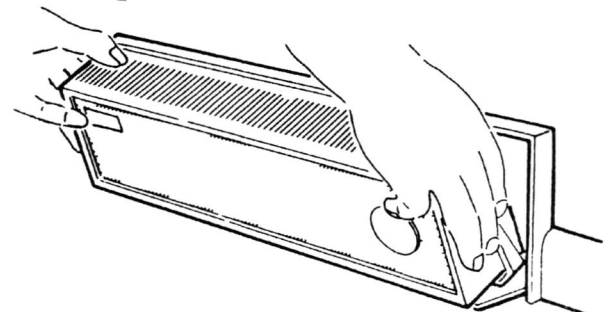

Fig. C-1.

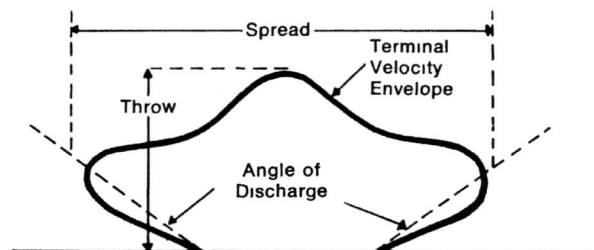

Fig. C-3.

though other location can be used. However, seasonal adjustments in the direction of flow from such other locations will probably be needed. For example, a supply outlet located low on an inside wall and discharging horizontally may perfom well on heating because of the natural tendency of the warm air rise and mix with room air. Such an outlet would be designed for a relatively short blow so that the jet stream would not normally be noticed by anyone in the room. However, this same location would not be satisfactory if the outlet were discharging cool air, since this air would not rise naturally to mix with room air. A seasonal adjustment of the outlet vanes would improve summer performance by establishing an upward angular projection that would result in rapid mixing with room air.

A supply outlet located high on an inside wall and blowing horizontally toward an outside wall will perform very satisfactorily for cooling provided the throw is less than the room width. If the throw is greater than the room width, the air stream will splash against the opposite wall and deflect downward as an undesirable draft in the occupied zone. Cool air blown into the room above head level will diffuse gently down into the room due to its greater density as compared with the room air. On the other hand, warm air discharging from the same location will tend to stay high due to its lighter-than-room-air density. The result will be poor mixing and stratification during the heating cycle and no counter-action to cold wall down drafts (as we have with low perimeter outlets). The ceiling diffuser has the same advantage for cooling and disadvantage for heating as has the high sidewall register. Here are some good rules of thumb for selecting outlets.

Floor diffusers, baseboard and low sidewall outlets:

Heating only	—select for 4 ft throw at design cfm and pressure limitations
For cooling	—select for 6 ft throw as design cfm and pressure limitations

Ceiling outlets:

360 or wide spread type	—select to throw equal to distance from outlet to nearest wall at design cfm and pressure limitations
Narrow spread type	—select for throw equal to ¾ of distance from outlet to nearest wall at design cfm and pressure limitations.

Two adjacent ceiling outlets	—select each so that throw equals ½ distance between them at design cfm and pressure limitations.
High sidewall outlets	—select for throw equal to ¾ of distance to nearest wall at design cfm and pressure limitations. If drop excessive, use several smaller outlets rather than one large outlet to reduce drop.

Placing Returns

Perhaps the most frequently asked question concerning systems design is "Where should the returns go?" This question seems to persist year after year despite the fact that there is nothing magical or mysterious about return air intakes. It may well be that the influence of returns in the overall room air distribution picture is considerably overestimated.

Essentially, a return air intake is nothing more than a relief valve that bleeds off unwanted air from a conditioned space. And, like all well-designed relief valves, except within the immediate vicinity of the inlet, the device causes little or no disturbance to the main stream—in our case, air in the occupied area of a room.

(Note: The occupied area of a room is considered to be the inner core of the space, physically extending 6 ft above the floor and to within 1 ft of all walls.)

Actually, for all practical purposes, it's the characteristics, number and placement of the *supply* outlets that determine whether room air distribution—that is, temperature uniformity and air motion—is good or bad. Generally, money spent on the supply system does far more good than equal dollars spent on the return side. (We assume that both sides of the system are properly *sized*.)

But since a return is a relief valve, it should be installed so that the unwanted air is drawn off first, say air that is too hot or too cold. In this way, returns can ease the burden on the supply outlets, which must mix conditioned air and room air together into a homogenous, draftless mixture.

Now in every room there exists a region referred to as the *stagnant zone* or stagnant layer. In this region, the supply air discharge ceases to have an influence or exert agitating action on room air. Room air merely stratifies in layers of descending temperature. We have all seen cigarette smoke "hang" in one spot in a room (at the floor, or near the ceiling). This is visual evidence of the existence of a stagnant zone.

The general rule for the placement of returns is to locate intake grilles *within* this stagnant zone. In this way, the least homogenous air will be drawn out of the room. During *heating,* the stagnant zone is near the floor, so returns should be placed *low* for both high and low supply outlet arrangement. During *cooling,* the stagnant zone is near the *ceiling,* and again, irrespective of the supply locations, returns should now be placed high.

In the case of a single system that must both heat and cool, returns should be located to serve the most critical season—the period when the most disturbing stagnant zone exists. Thus, with perimeter (low) supply, the larger stagnant zone occurs during cooling operation, so returns should be placed high. With high supply locations, the poorer year 'round performance occurs during heating, so returns should be placed low to bleed off air near the floor.

PUMP SELECTION

In general, pumps are used in solar heating to produce fluid flow and therefore serve the same basic funtion as fans—to transfer or move a fluid from one point to another. The main difference between the two is that pumps are intended to move a liquid—usually water or antifreeze—fans move air. And just as there are many kinds of fans there are many pump designs too. Reciprocating and centrifugal pumps are the most common but there are also gear pumps (used in oil burners), turbine pumps, and other special purpose devices. In this review we shall discuss, specifically, the centrifugal pump and its application to solar assisted heating systems, and provide answers to common pumping problems. However, an analysis such as this requires an understanding of the characteristics of the system and the pump. Therefore, a review of these points is in order.

The rotating wheel (impeller) of the centrifugal pump (see Figure D-1) causes water to flow from the pump inlet—which is open to the center of the wheel—to the pump outlet—which is open to the wheel periphery. The flow of water through the wheel at a given wheel speed is dependent upon the restriction to flow imposed on the wheel inlet and outlet. Quite logically, the wheel can handle the most water when no attachments are made either to the pump inlet or outlet. As restric-

tions to flow are imposed in the water passages, flow will diminish. These restrictions—the sum of which are called pump (or total) *head*—fall into four categories: *static head, friction head, pressure head,* and *velocity head*. Thus we can say:

$$H_t = h_s + h_f + h_p + h_v$$

where

H_t = Total head
h_s = static head
h_f = friction head
h_p = pressure head
h_v = velocity head

SYSTEM FLOW RESTRICTIONS

Static head (h_s) occurs when the pump must lift water against the force of gravity. To correctly estimate the amount of static head, a measurement must be made from the level at which water stands in the system when the pump is not running to the highest point that the water must be lifted. Using the view of the cooling tower in Figure D-2 as an example, we would measure from the water level in the tank (the level at which water stands in the system when the pump is not running) to the spray nozzle above the tower (the highest point to which the water must be lifted), which would give us a measurement of 7 feet. The temptation to ignore these levels and measure, instead, the height of piping must be resisted. Such a practice would have resulted in an incorrect measurement of 24 feet instead of 7 feet in the example cited above. In a drain-down type of solar system (see Lesson 9), the lift would again be the difference between the level that the water stands in the system and the highest point to which the water must be lifted. In the arrangement in Figure D-3 this difference would be 18 feet—from the water level in storage to the top of the collector.

Friction head (h_f) is a measure of the restriction to flow caused by the friction of water as it flows through the system. Unlike static head, friction head is directly related to flow rate and varies as the square of the flow ratio—(new flow/old flow)^2. Thus, if the flow is doubled, the friction head will be four times as great. The method of determining friction head is essentially the same as that used in determining pressure losses in duct systems, i.e., totaling the lengths of pipes, counting the number of fittings, valves, etc. and consulting charts in the ASHRAE Guide and other

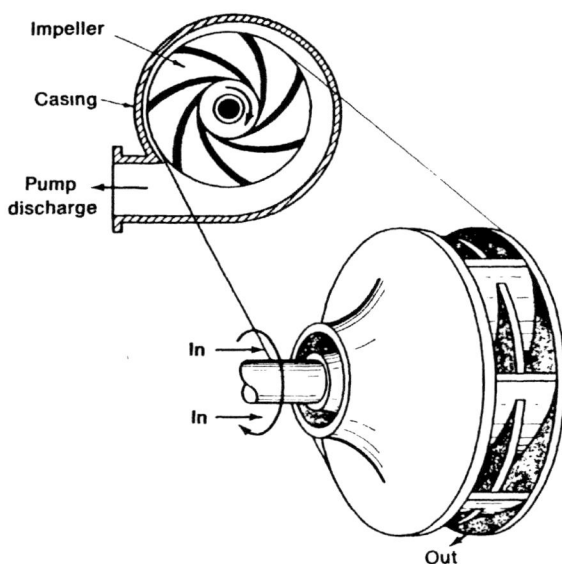

Fig. D-1.

Impeller
Casing
Pump discharge
In
In
Out

references for pressure loss or equivalent length values. In a closed solar system such as that shown in Figure D-4, friction head would be the only restriction to flow.

Pressure head (h_p) is a measure of the restriction of flow due to pressure maintained in the pipe system. Such a head would be present if it were required to pump water from a lower tank, which is open, to a closed tank maintained at a pressure above atmospheric by the use of compressed air. Pressure head, like static, is independent of flow.

Velocity head (h_v) is analogous to velocity pressure in a duct system. However, the value of the velocity head is usually very small in relation to the other components comprising the total head. As a result, it can usually be neglected in computing the total head against which a pump must operate.

Curves Represent System Characteristics

Graphical representation of pressure-flow interrelationship is an important tool in the analysis of pumping problems. Flow, in gallons of water per minute, is plotted on the horizontal axis. Head, in feet of water, is plotted on the vertical axis. (It should be noted here that a column of water 2.31 feet high exerts a pressure at its base of 1 lb per square inch. Thus, while we talk about pump head as feet of water, we should remember that this is merely a convenient way of expressing pressure.)

In the first curve diagram (Figure D-5) we can see how static, friction and pressure heads of a system are related to flow. Remembering that static pressure heads are independent of flow we can determine their value, say 25 feet, which stays constant as they remain the same regardless of flow. In the graph, this value is indicated as line X. Friction head—related to flow as the square of the flow ratio—appears as curve A. Note that at no flow the total system head consists of the static and pressure heads. At a flow of 100 gpm the friction head is say, 15 feet, making the total head 40 feet (25 + 15). At twice the flow or 200 gpm, the friction head is $(200/100)^2$ or four times the value at 100 gpm, thus 60 feet. Total head is 85 feet (25 + 60). If the piping system were such that there were no static or pressure heads, line X would be coincident with the 0 feet head line and the friction curve A would start at 0 instead of at the 25 feet head level. If the friction head at 100 gpm were more or less than the 15 feet assumed in the above example, curve A would fall above or below its plotted position in the illustration. It would, nevertheless, follow the relationship that the friction heat varies as the square of the flow ratio.

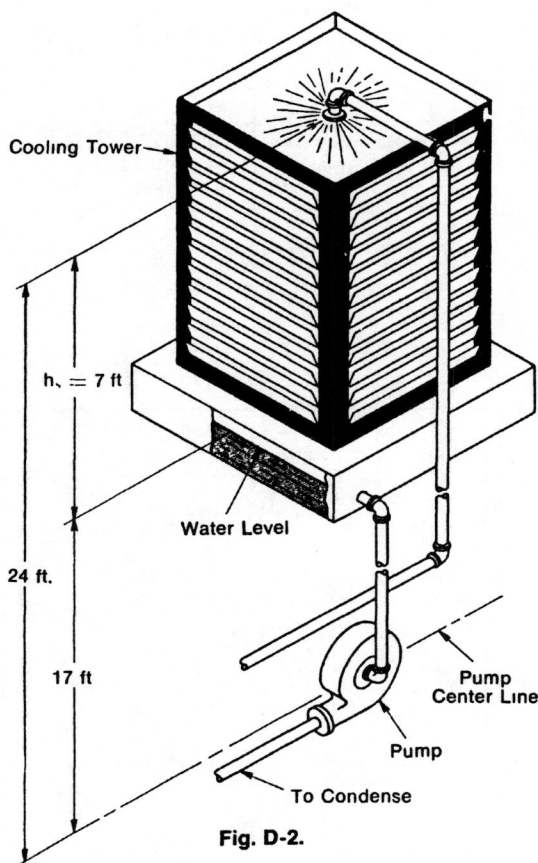

Fig. D-2.

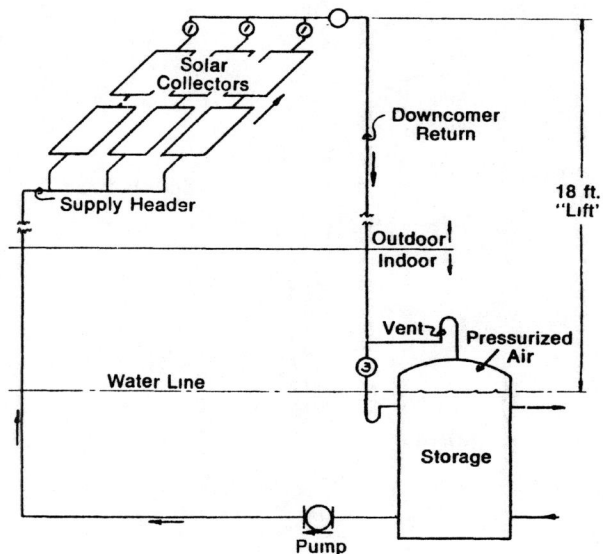

Fig. D-3.

The curves in Figure D-5 represent system characteristics only and have nothing to do with any pump. All they show is how much head (or pressure) is required to force water through the system at various rates of flow. To be expected, greater flow rates require greater pressures.

Combination Curves Aid Analysis

In the set of curves in Figure D-6, pump characteristics curves A and B are superimposed

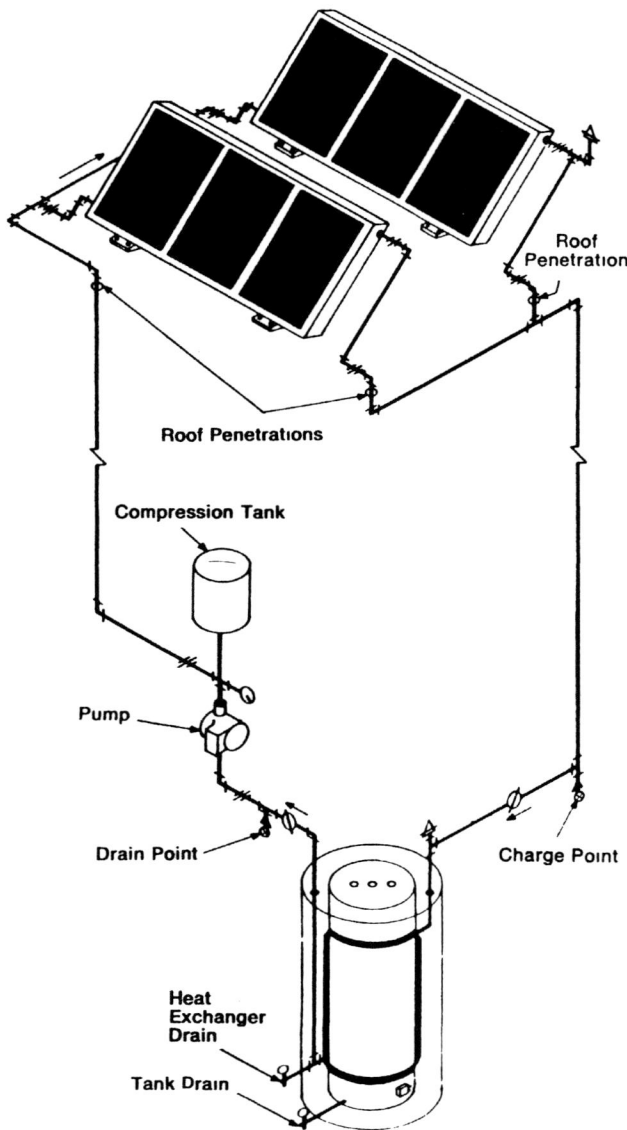

Fig. D-4.

on system characteristics curves X and Y (such as the one shown in the first curve diagram). By doing this we can readily see the results of changes in pump speed and/or system characteristics. From the pump characteristic curves we can see that the maximum head a pump can develop is read at the zero flow line. Naturally it would be inadvisable to operate a pump very long with its discharge valve closed—churning water would develop heat which might eventually damage the pump casing. However, in some instances, a pump is required to fill a high system and the pump head must be kept at near shut-off until the system is filled. In such cases, pump head at near shut-off is of significance. As the head or restriction to flow is reduced we can see from pump curves A and B that a pump can handle an increasing quantity of water. Pump curve A, which

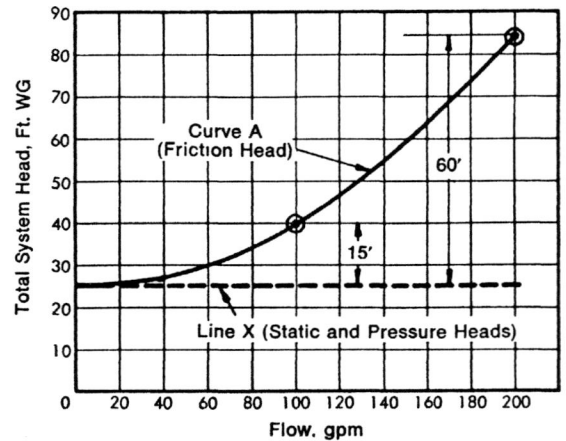

Fig. D-5.

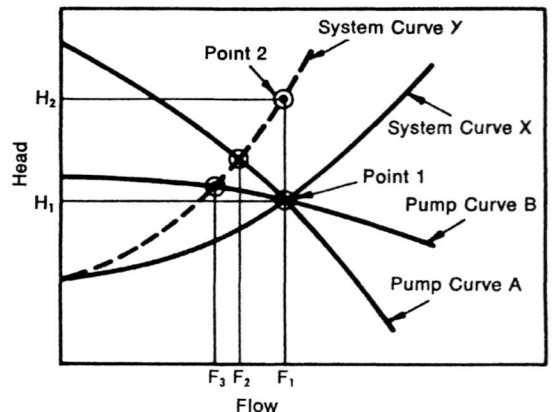

Fig. D-6.

falls off sharply, is known as a steep curve compared to flat curve B which tends to be more horizontal. The steep curve, which is characteristic of a pump having an impeller with backwardly inclined vanes, is preferable for air conditioning work. By using this type of pump, flow—which is of primary concern—is not affected as much by head variations.

To Find System Flow Potential

When we superimpose a pump curve on a system curve we can determine how much water will flow through a system by obtaining their points of intersection. For example, at the intersection of system characteristic curve X and pump characteristics curve A (Point 1) we read a flow of F 1 and a pump head of H 1. The balance point of a pump and system will always lie on both the pump and system curves since the two curves only intersect at one point.

To Select the Correct Pump

We can also use Figure D-6 in another way. Assume we have designed a piping system for a flow F 1 and have determined the total restriction to flow to be H 1. Using the relationships that have been discussed, we calculate the head for several other flow rates and establish system curve X. From a pump catalog we find two pumps that can meet the requirement for flow F 1 and head H 1, and then plot through this point, pump curves A and B.

As is often the case, suppose we had incorrectly estimated the system head at flow F 1. To show what would happen let's assume we underestimated and that instead of the head being H 1 it was actually H 2 at F 1 and the higher system characteristic curve Y resulted. It can be readily seen that neither pump A nor pump B can provide water at the Point 2 conditions of flow. Head-pump A balances system Y at flow F 2 and pump B balances system Y at flow F 3. Notice that the possible error in head calculation makes a smaller difference in flow with the steeper pump curve A than with the flatter curve B—bearing out our earlier statement. This would also be true had we overestimated system head.

Two Pumps vs One Pump

Figure D-7 showing pump-system balance for assumed conditions can provide the answer to this representative question: A certain pump will deliver 105 gpm through a piping system. How much water will be delivered if a second identical pump is installed to operate parallel with the original pump?

In this diagram, curve M represents the characteristic of a single pump. The capacity of two parallel running pumps is represented by the pump characteristic curve N. Points on this curve represent twice the flow for a given head as do the points on curve M. Thus, at a head of 30 feet, pump M can handle 150 gpm whereas the two pumps can handle 300 gpm.

Curve S represents the characteristic curve of the system to which the pumps are applied. Since the system curve shows a value of 30 feet at no flow, we know that this is the sum of the static and pressure heads. Curve S intersects curves M and N at Points 1 and 2. The value of flow at these two points are 105 and 135 gpm respectively.

Two Other Curves to Consider

Thus far we have discussed only one of the characteristic curves of a pump, the curve of flow vs. head. There are two others—horsepower and efficiency curves.

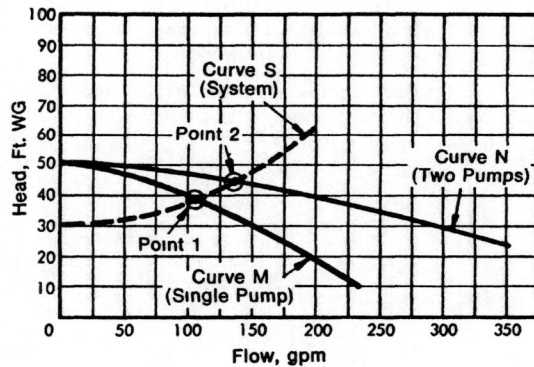

Fig. D-7.

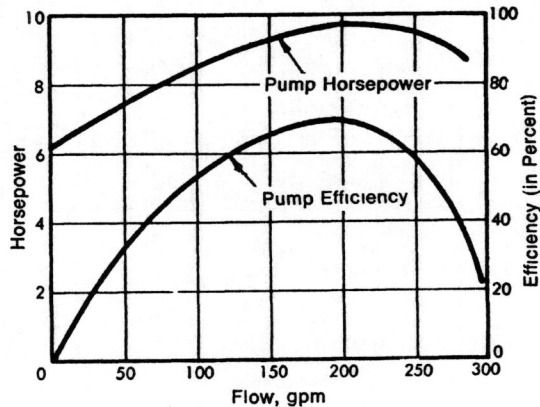

Fig. D-8.

Several things should be said about horsepower. One is that the actual horsepower required is always more than the theoretical amount due to frictional and other unavoidable losses. Brake horsepower can only be determined by an actual test of the pump. Another point of interest is the fact that pumps are usually designed with a self-limiting horsepower curve. As illustrated in Figure D-8, the horsepower requirement for a given pump at a given speed rises to a peak and then falls off as flow increases. If the motor is selected to handle this peak requirement, it cannot be overloaded by the increased flow that would result should the actual head be less than estimated when the pump was selected

The second curve in Figure D-8 (placed there for convenience and having no relationship to the horsepower curve above it) is a typical efficiency curve. Note that it too rises, then falls. Efficiency is calculated by divviding theoretical horsepower by actual test horsepower values. Thus:

$$H_e = \frac{h_t}{h_a}$$

where

H_e = Horsepower efficiency
h_t = theoretical horsepower
h_a = actual test horsepower

In our graphic example the efficiency curve shows that this particular pump would be satisfactory for use from 125 to 250 gpm where its efficiency ranges over 60 percent. Pumps having an efficiency lower than 60 percent in a given flow range can be used, but pump efficiency is a factor in operating cost and should be considered when there is a choice of pumps for a specific application.

SOME ADDITIONAL FACTS ABOUT PUMPS

Although pumps can be applied to discharge water at almost any desired head, care must be given to limitations which exist on the suction side of the pump when the water source is below the level of the pump. In this type of system, if a complete vacuum were drawn in a vertical pipe with its lower end in the water supply tank, the water would rise approximately 34 feet due to the weight of air on the surface of the water in the tank. If the pump could produce a perfect vacuum and if there were no losses, 34 feet is the maximum distance water could be lifted in the suction line.

When normal pipe losses and the necessary pressures needed to get water into the pump are considered, usual pump applications require that the suction lift be limited to 15 feet.

When hot water is to be handled, it is necessary to limit the negative pressure to a value above the pressure of vaporization. Otherwise the vapor formed will produce erratic pump operation.

In order for a centrifugal pump to deliver, its casing must be full of water. A number of methods can be used to assure this. The simplest is to install the pump below the lowest level of water entering the suction line. Another method is to employ a check valve in the suction line to prevent the initial prime from draining out when the pump is shut down. If neither of these methods is practical, it is necessary to provide an available water supply or to use a small hand pump. A vacuum pump is sometimes used to draws water up into a casing for priming.

Pipe Sizing

The resistance to flow, termed friction head (h_f), is made up of the resistance in straight lengths of pipes, elbows, tees, couplings, various hand valves and control valves, plus the solar panels, heat exchangers, and any other device through which liquid must be pumped.

Collector manufacturers usually provide the pressure loss (or resistance to flow) imposed by the collector for various flow rates through the device (see Figure D-9). In addition, collector manufacturers usually specific flow rate through panels (approximately .02 gpm per square foot). Other component manufacturers provide similar data.

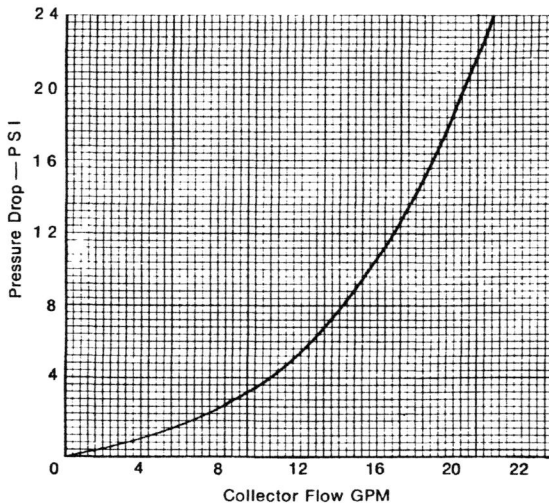

Fig. D-9.

The resistance or pressure drop in straight lengths of pipe can be obtained from basic friction charts in the *ASHRAE Handbook*. Pressure loss through elbows is listed in terms of the "equivalent length" of section of straight pipe; for example, a flow velocity of 5 feet per second, a four inch elbow imposes the same resistance to flow as 11 feet of pipe. In turn, other fittings are related to "equivalent elbows" in terms of pressure loss. Again, for example, an open globe valve is equivalent to twelve 4-inch elbows. Thus a four-inch globe valve imposes the same resistance as 12 × 11 or 132 feet of pipe.

To estimate the total (and greatest) resistance of a piping circuit, usually (but not always) the longest run is measured—both horizontal and vertical distances; the number of elbows and other fittings are counted and converted to equivalent feet of pipe, and then the measured and equivalent feet of pipe are added together to find the "effective" pipe length. Knowing the flow and total effective length, it is possible to find the pressure loss for any common pipe size by use of the ASHRAE friction charts. As in the case of ductwork, there are a number of simplified procedures to ease the work of design. Table D-1 provides one example.

Table D-1 assumes a standard number of fittings in a circuit, and includes the losses in the tabulated data so that the designer merely mea-

TABLE D-1
PIPE SIZING-HEAD PRESSURE TABLE

AVAILABLE HEAD ft. of water	TOTAL LENGTH OF CIRCUIT (AS MEASURED ON PIPING LAYOUT)																	
	a	b	c	d	e	f	g	h	i	j	k	l	m	n	o	p	q	r
4	35	45	50	60	65	70	75	80	90	100	110	130	150	180	220	290	400	620
5	45	60	65	70	80	90	100	100	120	130	140	160	190	230	290	360	510	790
6	55	70	80	90	100	110	120	130	140	160	180	200	240	290	350	450	620	950
7	65	90	100	110	120	130	140	150	170	190	210	240	290	340	420	540	730	1120
8	75	100	110	130	140	150	160	180	200	220	250	290	330	400	490	620	850	
9	85	110	130	150	160	170	190	200	230	250	290	330	380	450	560	710	950	
10	100	130	140	170	180	190	210	230	260	290	320	370	430	510	620	790	1060	
11	110	140	160	190	200	220	240	260	290	320	360	410	480	570	690	880	1170	
12	120	160	180	200	220	240	260	290	320	350	400	450	530	620	760	960		
14	150	190	210	250	260	290	310	340	380	420	470	540	620	730	900	1120		
16	170	220	250	290	310	330	360	400	440	490	550	620	720	850	1020			
18	190	250	290	330	350	380	420	450	500	560	620	710	830	950	1150			
20	220	290	320	370	400	430	470	510	560	620	700	790	910	1060				

PIPE SIZE	GALLON PER MINUTE CAPACITIES																	
⅜"	09	08	07	07	06	06	06	06	0.5	05	05	05	04	04	03	03	0.3	02
½"	2.3	2.0	19	1.8	1.7	17	16	15	1.5	14	13	12	12	11	0.9	08	07	06
¾"	5.0	43	4.1	38	37	36	3.4	3.2	3.1	29	28	26	2.4	22	20	18	1.6	1.3
1"	96	8.3	7.7	73	70	68	65	6.3	5.9	57	55	50	46	43	38	34	29	23
1¼"	—	18	17	16	155	15	14.5	14.0	13.0	120	115	110	97	90	83	73	6.3	4.8
1½"	—	27	25	24	23	22	21	20	19	18	17	16	15	13	12	11	93	7.5
2"	—	—	—	—	—	—	42	40	39	36	34	32	29	27	24	21	18	14
2½"	—	—	—	—	—	—	—	—	—	57	54	52	47	44	38	32	29	24

* ⅜" Copper Tubing only.

NOTE. Do not go beyond the maximum or below the minimum figures shown in the table.

HOW TO USE THIS TABLE FOR FINAL PIPE SIZE SELECTION

(a) Single Pump

Enter the upper portion of the Table at the head pressure of the pump selected. Read across to the figure closest to the total length of circuit. Read down to the lower portion of the Table to the gpm figure equal to or greater than the gpm required for the circuit. Read to the left-hand column to determine the pipe size required. Repeat for each circuit. Staying in the same column established by the circuit with the longest total length, repeat the last step for the gpm requirements of the trunk and distribution piping.

(b) Multiple Pumps

Enter the upper portion of the Table at the head pressure of the pump selected. Read across to the figure closest to the total length of the longest circuit served by the pump. Read down to the lower portion of the Table to the gpm figure equal to or greater than the gpm required for the circuit. Read to the left-hand column to determine the pipe size required. For a two-pipe circuit, size all piping in the circuit from the same column in the Table established above.

Size the trunk and any distribution piping using the total gpm of the system, the lowest head pressure of the pumps selected, and the longest total length of circuit.

sure the length of pipe—say from the pump to the collectors and back, then knowing the available pump head and gpm flow rate, a proper pipe size is determined. For example, assuming a pump has a 12 foot head, enter the table at the row showing 12 foot head. If the measured length of the circuit was 180 feet, move right and stop at the number, then move down until the flow rate, let's say 7 gpm, in the circuit is reached (or exceeded). Follow the line to the left and note that a 1 inch pipe size is required. (Broken lines and arrows trace the pattern.)

Again, as with any simplified sizing charts or tables, be certain you understand the assumptions used to develop the procedure and the limits of its application to your piping arrangementa.

For additional pipe sizing information, contact: Hydronics Institute, 35 Russo Place, Berkeley Heights, New Jersey 07922.

SOLCOST
Solar Service Hot Water Form

- Refer to the **SOLCOST** Solar Hot Water Handbook for instructions.

If you do not have the necessary experience in heating, plumbing, and/or hot water systems, consult your contractor, engineer, solar systems manufacturer, supplier, or utility company for assistance in completing this form.

- The completed form should be mailed to:

Solar Environmental Engineering Co., Inc.
SOLCOST Service Center
P.O. Box 1914
Fort Collins, Colorado 80522   (303) 221-4370

Service charges and **SOLCOST** sales information may be obtained from the
SOLCOST Service Center listed above.

- Items on this form marked * have values that will automatically be used if the user does not make an entry. Refer to Default Table on back page of form for default values.

** Important — Refer to Handbook for Guidance **

Name _____   Phone Number _____ Date _____

Address _____   Contractor/Estimator/Designer Name _____

_____

_____ Zip _____   Phone Number _____

A. SOLCOST Analysis Description

1 This SOLCOST analysis is for a ____ owner occupied residential building

____ business related, rental building, or commercial building

____ non-profit organization owner (i e., public buildings, schools, etc)

2 SOLCOST determines the optimum solar collector area which maximizes the rate of return (or present worth) of the solar investment
Do you want SOLCOST to optimize your collector size? ____ yes ____ no

3 If the previous answer is no, you must enter collector area in square feet _____

4 SOLCOST also determines the optimum size for heat exchangers, pumps or blowers, and pipes or ducting.

B. Solar and Conventional System Description

1 Building location _____
city, state

2 Building type _____ one or two-family residential
_____ multi-family residential
_____ commercial

3 Application ____ retrofit
____ new construction

4. Energy source for a conventional service hot water system which would be installed, if solar were not feasible
____ natural gas   ____ fuel oil   ____ propane
____ electricity   ____ coal

5 Auxiliary fuel for service hot water system (fuel type to be used in addition to solar)

____ natural gas   ____ fuel oil   ____ propane
____ electricity   ____ coal

6 If more than one fuel type is to be considered, enter a (2) for the alternate fuel   A complete SOLCOST analysis will be made for each fuel

** Important — Refer to Handbook for Guidance **

C. Solar Collector Subsystem Description

1. Collector Orientation

a) Azimuth Angle * _____ degrees

(0 is South, East = (+) degrees, West = (-) degrees)

The azimuth angle represents the direction the solar collector faces, usually due South

b) Tilt Angle * _____ degrees

(0 is horizontal, 90 is vertical)

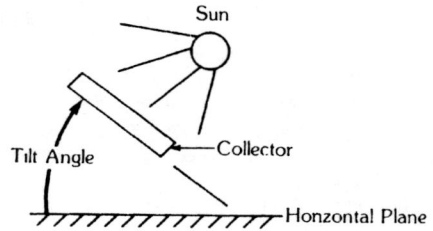

2. Collector Type

_____ air   _____ liquid

3. Collector Efficiency

This section must be completed with data obtained from the solar systems manufacturer, supplier, or contractor under consideration

Fill in *one* of the following

- FRPRIME-UL product _____ BTU/Hr-Ft^2-^{0}F and FRPRIME Tau-Alpha product _____

or - Efficiency at $(T_{in} - T_{amb})/I = 0\ 0$, _____ (efficiency) and

Efficiency at $(T_{in} - T_{amb})/I = 0\ 5$, _____ (efficiency)

or - Manufacturer _____

Address_____

Model Number _____

4. Expected life of solar collector * _____ years.

D. Solar and Conventional System Configuration

In the Handbook are three system configurations that are commonly used in residential and commercial solar hot water applications

SOLCOST uses these systems as a model' when calculating optimum sizes and costs

For one and two family residential applications, calculations will be based on the configuration A

For multi-family and commercial applications, SOLCOST will automatically select configuration B or C, depending on optimal performance and cost

The following information is required regardless of application

1. Solar storage tank

What is the size of the solar storage tank, if known _____ (gallons)

(If solar storage tank size is not known, SOLCOST will provide a reasonable size estimate)

2. Solar hot water piping (ducting) description

Length of pipe (or ducting) required between collector and solar storage tank _____ (feet, total length of supply and return)

3 Solar hot water freeze protection (liquid collectors only — check one)

Anti-freeze _____   Drain-down _____   None _____ (Tropical climate only)

E. Hot Water Loads

1. Residential Application

Fill in *one* of the following:

- BTUs per day required for hot water _____

or - Gallons of hot water used per day _____ (gallons/day)

Number of occupants _____

Hot water set temperature _____ (^{0}F)

or - Estimated Consumption - using fuel bills for summer water heating

Hot water delivery temperature (if known) _____ ^{0}F

Quantity of fuel consumed _____

(kwh, gallons, tons, 100 cu ft , etc)

Total fuel bill for period of consumption $_____

Number of days in consumption period _____

(period when furnace was turned off)

Note: Not applicable for electric water heaters unless on separate meter

2. Commercial Application

Fill in *one* of the following:

- BTUs per day required for hot water _____

(Include losses for circulation loop, if applicable)

or - Monthly fuel bill for hot water *only* for the past 12 months

$_____, $_____, $_____, $_____,

$_____, $_____, $_____, $_____,

$_____, $_____, $_____, $_____,

Fuel type _____

Fuel price per unit (actual) $_____

or - Gallons of hot water used per day _____ (gallons/day)

Hot water heater set temperature _____ (^{0}F)

Water main temperature _____ (^{0}F)

If circulation loop is involved, estimate losses in BTUs per day _____

F. Reference System Cost

1 Reference (conventional) system, initial installed cost * $_____

2 Reference (conventional) system annual maintenance cost $_____ (average annual maintenance costs)

Annual maintenance may be expressed as percent of initial installed cost * _____ % (Do not use for retrofit situations)

G. Solar System Cost

1 Solar Component Cost Information

a) Collector cost/square foot $_____/sq ft
b) Solar storage tank cost/gallon $_____/gallon
c) Controls $_____
d) Other $_____

2 Installation Costs of Solar System

a) Design/engineering costs $_____
b) Modifications to building and existing system
  - labor costs $_____
c) Installation labor costs $_____
d) Misc materials cost (paint, wallboard, shingles, etc) $_____

Total Installation $_____

3. Solar System Maintenance Cost

(Calculate *average annual* maintenance cost over life of system)• $_____
Annual maintenance may be entered as percent of solar system cost •   _____%

H. Finance and Tax Data

1. Residential

a) Loan interest rate •   _____%
b) Loan term •   _____ years
c) Loan down payment • $_____
d) Property insurance rate •   _____%
e) Property tax rate   _____%
f) Personal income tax rate   _____%

2. Commercial

a) Loan interest rate •   _____%
b) Loan term •   _____ years
c) Loan down payment • $_____
d) Property insurance rate •   _____%
e) Property tax rate   _____%
f) Corporate tax rate or
  owner income tax rate   _____%

3. If business application (commercial or residential rental, for example) check _____ and fill out the following:

a) Depreciation method (Options are straight line or declining balance)   _____
b) Multiplier used in declining balance depreciation •
  (limited to 1 5 for commercial buildings and 2 0 for new residential property)   _____
c) System useful life for depreciation purposes •
  (currently 10 years is allowed for building heating, air conditioning, and service hot
  water systems, except storage tanks, which are allowed 22 years)   _____ years
d) Salvage value of the solar system at the end of its useful life •   _____ % of total cost

I. Fuel and Electricity Cost and Price Escalation Data

Enter current cost for fuel, including local sales tax and fuel cost adjustment factors

1. Natural Gas Cost Schedule

BTU content per cubic foot • _____

Fuel Cost ($/cubic ft) *Quantity* (no of units)
Step 1 _____   for first _____
Step 2 _____   for next _____
Step 3 _____   for next _____
Step 4 _____   for next _____
Step 5 _____   for next _____

3. Fuel Oil Cost Schedule

Fuel oil grade • _____

Fuel Cost ($/gallon) *Quantity* (no of units)
Step 1 _____   for first _____
Step 2 _____   for next _____
Step 3 _____   for next _____
Step 4 _____   for next _____
Step 5 _____   for next _____

2. Electricity Cost Schedule

Fuel Cost ($/kwh) *Quantity* (no of units)
Step 1 _____   for first _____
Step 2 _____   for next _____
Step 3 _____   for next _____
Step 4 _____   for next _____
Step 5 _____   for next _____
Step 6 _____   for next _____
Step 7 _____   for next _____

4. Propane Cost Schedule

Fuel Cost ($/gallon) *Quantity* (no of units)
Step 1 _____   for first _____
Step 2 _____   for next _____
Step 3 _____   for next _____
Step 4 _____   for next _____
Step 5 _____   for next _____

5. Coal Cost Schedule

BTU content per ton * _____

Fuel Cost ($/ton)   *Quantity* (no of units)

Step 1 _____   for first _____

Step 2 _____   for next _____

Step 3 _____   for next _____

6. For the fuel(s) priced in 1 through 5 above, you may estimate the annual price escalation over the life of the solar system (collector life Default values are shown in the Default Table

      a) Estimated average annual escalation rate * _____ %   for conventional (reference fuel)

      b) Estimated average annual escalation rate * _____ %   for auxiliary fuel

---

Default Values

Whenever data in this form is asterisked (*), it means users either accept the data already in SOLCOST, i e , the Default Value, by not entering any information, or enter another value of their choosing Listed below, by section, are the SOLCOST Default Values

Section C — Solar Collection Subsystem Description

1 a) Azimuth Angle   0

   b) Tilt Angle   Latitude

4 Expected life for solar collector   20 years

Section F — Reference System Cost

1 Initial Installed Cost $0 00 (assumes retrofit, with cost already absorbed)

2 Maintenance Cost   01 (1% of the initial installed cost per year)

Section G — Solar System Cost

3 Maintenance Cost   01 (1% of initial installed cost per year)

Section H — Finance and Tax Data

1 Residential	New Construction	Retrofit
a) Loan interest rate	09 (9%)	11 (11%)
b) Loan term	20 years	7 years
c) Loan down payment	10 (10% of total loan)	20 (20% of total loan)
d) Property insurance rate	005 (5% per year)	005 (5% per year)
2 Commercial	New Construction	Retrofit
a) Loan interest rate	09 (9%)	11 (11%)
b) Loan term	20 years	7 years
c) Loan down payment	10 (10% of total loan)	20 (20% of total loan)
d) Property insurance rate	005 (5% per year)	005 (5% per year)

3) Business only

   b) Multiplier used in declining balance   1 25

   c) System useful life for depreciation purposes   10 years

   d) Salvage value of system at end of useful life   10 (10% of total cost)

Section I — Fuel and Electricity Cost and Price Escalation Data

1 Natural Gas   Heat Content per Unit   100,000 BTUs/100 cubic feet

3 Fuel Oil   Heat Content per Unit   139,600 BTUs/gallon (no 2 fuel oil)

5 Coal   Heat Content per Unit 30×10^6 BTUs/ton

6 Average Annual Escalation Rate

The following estimated escalation rates, through the year 2000, are provided from ERDA research data

Natural gas	8%
Electricity	6%
Fuel oil	7%
Propane	8%
Coal	6%